2·18·80

FIRE! HOW DO THEY FIGHT IT?

WESTMINSTER PRESS BOOKS
BY
ANABEL DEAN

Fire! How Do They Fight It?

Submerge!
The Story of Divers and Their Crafts

FIRE! HOW DO THEY FIGHT IT?

BY

ANABEL DEAN

THE WESTMINSTER PRESS
PHILADELPHIA

BOOK DESIGN BY DOROTHY ALDEN SMITH

First edition

Published by The Westminster Press ®
Philadelphia, Pennsylvania

PRINTED IN THE UNITED STATES OF AMERICA

1 2 3 4 5 6 7 8 9

ACKNOWLEDGMENT

The author and publisher wish to thank the Los Angeles Fire Department for furnishing information and photographs.

Library of Congress Cataloging in Publication Data

Dean, Anabel.
Fire! How do they fight it?

Bibliography: p.
Includes index.
SUMMARY: Discusses the chemistry of fire, techniques and equipment used to combat different types of fires, and various activities of fire departments.
1. Fire extinction—Juvenile literature. 2. Fire departments—Juvenile literature. 3. Fire—Juvenile literature. [1. Fire extinction. 2. Fire departments. 3. Fire] I. Title.
TH9148.D4 628.9'25 77–17635
ISBN 0–664–32626–9

CONTENTS

1
THE DEVELOPMENT OF FIRE DEPARTMENTS

WHAT MIGHT HAVE BEEN

One dark night thousands of years ago a group of people huddled at the mouth of a cave. They watched fearfully as lightning streaked across the sky. The little valley rumbled with thunder. It was late in the year and the cave people shivered in the cold wind.

Crash! A bolt of lightning struck a dead pine tree near the mouth of their cave. Thunder followed so closely it drowned out the screams of the terrified people. Now the little valley was lighted by an eerie glow as the pine tree blazed. The people gazed in terror but were afraid to flee past the fire. They drew fearfully back into the cave.

By morning the weather had turned colder. A few snowflakes drifted down. The small band of people shivered as they crawled out of their damp cave. The children whimpered and begged for food, but hunting had been poor and there was none to give them.

The top of the pine tree, which now lay smoldering, didn't seem so terrifying in the daylight. A few of the braver people drew closer to stare curiously. A wonderful warmth, similar to sunshine, came to them from the fallen log. Those near the log beckoned to the others and soon all stood warming themselves in the frosty air.

"Ummmmmm!" said one small boy, rubbing his stomach. The other hungry people too had begun to notice the delicious odor. They all glanced curiously around.

"Ut." A man pointed to the charred body of a deer at the foot of the pine tree. It must have been a victim of the lightning strike the night before.

The hungry people, mouths watering, gathered around the deer. What would happen to them if they ate burned meat? One man reached out and touched the animal. It was still hot. He jerked his fingers back and stuck them into his mouth. A look of surprise came over his face. He smacked his lips and began to pick pieces of the hot cooked flesh from the deer. The sight and smell were too much for the other hungry people. They began to tear off pieces of meat and cram them into their mouths. This was the best food they had ever eaten.

When the well-fed people straggled back to their cave, one man carried a branch from the smoldering log. Perhaps it would be possible to keep a fire burning at the mouth of their cave for warmth. They knew animals ran from fire. Maybe it would keep them away. A few were already wondering whether it would be possible to use fire to burn the next animal they killed.

We will never know exactly when or where fire was first used by humans. But it probably happened something like this story.

Peking men, people who lived in China about 500,000 years ago, were using fire. They must have obtained it from lightning-caused fires. In time, humans learned how to keep coals alive so they could start a new fire. Some groups learned to start a fire from the spark given off when a piece of flint was struck against pyrite. Others learned that the friction caused by rubbing two dry sticks together would produce heat to start a fire.

When people began using fire, they gained a valuable friend. But they also brought an enemy into their homes. A controlled fire is a great aid and comfort. An uncontrolled fire is a destructive force. When people began to gather into towns, a fire out of control in one dwelling could destroy the whole town.

EARLY FIRE FIGHTING IN AMERICA

The first settlers in America were frightened by the strange Indians, strange animals, and endless forests. They missed the safe established villages they had left in Europe. They built their homes in small clearings close together for protection and company. Almost everything they used was provided by the forests around them. Houses were built of wood with roofs thatched with straw or covered with wooden shingles. The outside of the chimneys was of wood and the inside, or flue, was plastered with mud or clay. After a few years, the roofing became dry and some of the clay lining the flue chipped off, leaving wood exposed to the heat.

All cooking and heating was done by wood fires in huge fireplaces. Flammable creosote and soot collect in chimneys where wood is burned. If the creosote caught fire, showers of sparks were shot skyward to fall on nearby roofs. Needless to say, there were many fires.

Everyone feared fires, which could quickly spread to all the houses. Almost every household had a leather three-gallon fire bucket with the family name painted on the side. At the dread cry of "Fire!" everyone dropped whatever he was doing, grabbed the bucket, and ran to fight the fire. If a fire broke out during the night, men running to the fire cried, "Throw out your buckets!" All over town windows were thrown open and buckets were tossed out. People picked them up and ran with them to the fire.

Everyone big enough to handle a bucket helped. People quickly formed a bucket brigade of two lines stretching from the fire to the nearest source of water. They filled buckets with water and handed them down the line for the water to be thrown on the fire. The empty buckets were passed back down the other line.

Getting enough water quickly was a problem unless there was a stream nearby. Often the only source of water was the town pump. When the distance was too far for the bucket brigade to reach, buckets were filled and hung from a pole. These were carried to the fire on the shoulders of two men.

There were other simple tools used in fighting fires. A long stick with a wet cloth swab at one end was used to put out fires in roof thatch. An iron hook on a rope, called a fire hook, was thrown up on a roof to catch in the burning thatch so it could be pulled down. A long ladder was used to reach the roofs and attics. A barrel of water was often kept handy for such emergencies.

EARLY FIRE PREVENTION LAWS

One cold night in January, 1653, a fire broke out in the seaport town of Boston. As was the case in all settlements then, there was no public water supply, fire department, watchman or other alarm system. The fire spread rapidly in the freezing weather while everyone slept.

When the fire was discovered, sparks and firebrands were raining down over most of Boston. No one was in charge of fighting fires, so people ran around without knowing what to do. Very little water was available—most of it was frozen. At this time, officials believed that all gunpowder should not be stored in one place, so barrels of gunpowder were stored all over town. Whenever fire reached a barrel of gunpowder, a loud explosion scattered burning material over a still wider area. Most of the terrified people tried to save their own household furnishings rather than fight the fire. One third of Boston burned down that night.

After this terrible fire, the officials of Boston held a meeting to decide how to prevent such fires. They passed these fire laws:

1. Every house must have a long ladder and a twelve-foot pole with a swab on the end.
2. Six long ladders and four fire hooks would be hung on the meeting-house to be used only for fires.
3. A public cistern was to be built in the center of town and fire buckets would be kept there.
4. No one was allowed to build an open fire within three rods of a building.
5. A night watchman would patrol the streets watching for fires.
6. No fires would be allowed on the ships tied up at the docks.
7. The penalty for arson was to be death.
8. No one could carry live coals in an open vessel.

This last law may seem odd, but it was very necessary. At this time, anyone who carelessly let his fire go out ran to the neighbors' for a few hot coals. These coals were usually carried home in an open container. When it was windy, sparks were often blown around.

Boston's fire laws cut down on the number of fires. But when a fire had a good start, there was little anyone could do.

THE FIRST FIRE DEPARTMENTS

As Boston and the other towns in America grew, bigger and finer buildings were built. Since there was no fire insurance then, fires made families homeless and put merchants out of business. All cities were plagued by terrible fires. In spite of Boston's fire prevention laws, a fire in 1676 destroyed forty-six houses, several warehouses, and Boston's North Meeting House.

The town officials of Boston heard about a "fire engine" being manufactured in England. Some of these engines were used during the Great Fire of London in 1666. The citizens of Boston were willing to try anything. In 1678 a fire engine was ordered from England.

The first fire engine, also known as a water engine or hand tub, looked nothing like a modern fire engine. It was a tub with a handle on top. It was carried to fires and placed in front of the burning building. A bucket brigade was formed to pour water into the single pipelike chamber. A piston, a disk

This kind of "fire engine," or hand tub, was used about 1765. The buckets are leather! NORTH CAROLINA MUSEUM OF HISTORY

that fit tightly inside the cylinder, was fastened by a rod to the handle. When the handle was raised, water could be poured into the chamber. When the handle was pushed down, the water was squirted out of the cylinder through a nozzle. But the uneven stream of water started and stopped every time the tub was filled and emptied.

A later-model water engine had two cylinders connected by valves to an air chamber in the center. A piston in each cylinder, connected to the long handle, was forced up and down as two men pumped. The water now did not squirt directly from the cylinder. It was pumped into the air chamber, where it compressed the air and built up pressure. The pressure pushed on the water, squirting it out of the nozzle in a steady stream as one cylinder sucked in water while the other pumped it out. When two men worked at top speed, enough pressure was formed to throw a column of water many feet.

In 1717 twenty of the leading citizens of Boston organized themselves into an unpaid volunteer fire department. Each man ran to all fires carrying two fire buckets, a sack, a screwdriver, and a bed key. The tools were used to take the bed frame apart so it could be carried outside. The sack was filled with household goods and taken outside where one man guarded the valuables. This was the first fire department in the American colonies.

By 1720, Boston had purchased six of the new water engines. Since this equipment was of little use without trained men to operate it, the town was divided into fire districts. A volunteer fire department, called a Mutual Fire Society, was organized in each district under a fire warden.

Other towns in the colonies began organizing fire departments patterned after the ones in Boston. In 1731, New York ordered two fire engines from England. About 1736, Benjamin Franklin helped organize the first fire-fighting society in Philadelphia. But for another hundred years all fire fighting was done by volunteers.

IMPROVEMENTS IN FIRE-FIGHTING EQUIPMENT

At first men ran to the fires carrying hand tubs. But later, wheels were added so they could be pulled by long ropes. Then the fire engines could be made larger.

It took months or years to order and receive a fire engine from England. Americans tried making their own fire engines and soon improved them. The first hand pumper built in the colonies had a gooseneck pipe with a nozzle that could be swiveled to throw a stream of water in any direction. All the first pumpers had only a nozzle attached to the air chamber. The Americans added a long hose so that the pumpers did not have to be operated so close to the fire. When the value of longer fire hoses was seen, a separate cart for the reels of fire hose was added. This cart, which was pulled to every fire, was the first piece of equipment with a bell to warn people of its approach.

The next improvement was a suction pipe to transport water from a stream, a well, or a cistern to the pumper. This eliminated the need for a bucket brigade.

The larger pumpers had long handles on both sides so fifteen or more men could pump at once. Pumping was done at man-killing speed to provide a strong jet of water. Usually two teams worked in shifts, since one team was exhausted after twenty minutes of hard pumping.

There was great rivalry between different teams and fire companies. Each one wanted to get to the fire first and put it out. Men from one team would sometimes slash the hose or upset the pumper of a rival company to prevent them from putting out a fire. At times, fights broke out as several fire companies tried to race to a fire, lay their lines, and produce the strongest stream of water.

After a few years, two types of hand pumpers had developed. The New York style had two long handles that ran the length of each side of the pumper. The men stood on either side of the pumper as they pumped. The air chamber was set over the rear wheels. The Philadelphia style pumper, also known as a double-decker, had the air chamber in the center. Two lines of men operated the handles on both ends of the pumper. One line stood on the ground while the other line stood on a platform that folded out from the engine.

One of the first fire engines equipped with wheels—pulled and pumped by manpower

PICTURE COLLECTION,
NEW YORK PUBLIC LIBRARIES

At first long ladders were kept in a few central locations to be brought to the fires as needed. Later, a long cart was used to carry ladders, buckets, and other equipment. This hook-and-ladder cart was pulled to every fire.

One hundred or more men were required by a fire company to pull the equipment to the fires and operate the pumpers. These volunteers performed an important service for the town. But they also enjoyed the excitement of running to a fire and the rivalry of trying to outdo each other. The fire companies were also social clubs. They held picnics, dances, and other events for the members and their families.

The first fire companies were made up of the town's leading citizens who were interested in protecting the town from fire. But unfortunately, as fire companies grew larger and larger, many thugs and bullies were attracted to the excitement of fire fighting. A barrel of rum was present at every fire of any size to refresh the workers. Since the men were unpaid, many felt they didn't have to take orders from the fire warden in charge. The conduct of the fire fighters at some fires got out of hand.

In 1832 a cholera epidemic swept New York City. There were not enough healthy men to pull the fire equipment to the fires. Horses had to be used. Fire officials were impressed by the speed and efficiency with which the horse-drawn equipment moved to a fire. Word of this spread across the United States. Many citizens thought horses would be a good addition to the fire departments. But the volunteers fought this move. They liked the excitement of fighting a fire and wanted to pull the equipment themselves.

By the early 1800's, steam engines were being used to operate steamboats and locomotives. In 1829, George Braithwaite of London built the first fire engine which used steam to pump water. Since it weighed over two tons, it was pulled by horses. Braithwaite built five steam fire engines in all. But the people of London preferred their hand tubs, and steam fire engines were not adopted there at that time.

Fires in the growing city of New York alarmed the city officials. In 1841 they decided to build one of the new steam fire engines. Although this engine was a great success in trials, volunteers refused to use it or allow it to be used at fires. The fire companies by now were too powerful. The New York City officials had to give up their idea of using steam fire engines.

It was years before steam engines and horses were accepted as part of every fire department. In many cities volunteer fire fighters put up a fight when steam fire engines were first introduced. Citizens became enraged when the volunteers fought among themselves to see who would get to put out a fire while their city burned.

New York style hand pumper. The tank still had to be filled from buckets

AMERICAN LAFRANCE CORP.

Philadelphia style hand pumper used around 1840. Bars flip out into round clamps and men stand on hinged platforms to pump

MUSEUM OF THE CITY OF NEW YORK

"The Night Alarm," from The Life of a Fireman, 1854. Lantern carrier and horn blower run ahead. Published by N. Currier
THE HARRY T. PETERS COLLECTION,
MUSEUM OF THE CITY OF NEW YORK

Cincinnati, Ohio, ordered one of the new steam fire engines around 1850. The city officials hoped it would put an end to the disorderly fights among the volunteers at every fire. But at the first big fire, the volunteers attacked the steam engine with rocks and clubs instead of trying to put out the fire. The angry citizens, seeing their homes burn while the firemen fought among themselves, attacked the volunteer firemen. There was a tremendous battle and the firemen were defeated. The steam fire engine, operated by a few citizens, threw four streams of water at one time and put out the fire without any help from the volunteers. This was the end of hand tubs in Cincinnati. It was the first city to replace the volunteers with horse-drawn steam fire engines.

By 1900, paid firemen with steam fire engines and trained fire horses

"The New Era of Steam and Muscle," from The Life of a Fireman. Engines are still pulled by men. Murray and Church Streets, New York City. Published by Currier & Ives in 1861

THE HARRY T. PETERS COLLECTION,
MUSEUM OF THE CITY OF NEW YORK

Team fire engine built around 1890. The driver rode, but firemen still ran alongside

AMERICAN LAFRANCE CORP.

Steam fire engine pulled by matched horses. San Francisco Fire Department, 1897

WELLS FARGO BANK HISTORY ROOM, SAN FRANCISCO

were used in all cities. Only trained men could operate and repair these engines. Even small towns, although depending on some volunteers, usually had one steam fire engine.

A steam fire engine with three matched horses was an exciting sight as it raced to a fire. At the sound of the fire gong, the trained fire horses ran to their places. A trained team could be harnessed and ready to roll in twelve seconds. The horses strained together as they pulled the clattering, clanging, smoke-belching pumper through the streets.

At the fire the pumper and other pieces of equipment were wheeled into place. The horses were unharnessed and led to a safe spot. Those horses were well cared for. The loss of a horseshoe could prevent a piece of equipment from reaching a fire.

Today all fire departments use motorized equipment, and some of the romance has vanished from fire fighting. The day of the hand tub and the horse-drawn steam engine is gone forever. But when the sirens wail and the red lights flash, we still feel the excitement of fire fighting as we watch the swaying firemen cling to the lurching trucks while they speed toward a red glow in the sky.

2
THE FIRE TRIANGLE

WHAT IS FIRE?

It is after dinner in Yellowstone National Park. Campfires flicker as tired campers gather around their welcoming glow. It grows darker. The sounds of strumming guitars and singing voices blend with the pop of burning wood. But most of the campers are quiet, content to just gaze into their campfires. They are fascinated by the changing colors and shapes of the flames. There is a mystery there. What process changes wood into ashes while giving off heat and light? The same thoughts must have been in the minds of the first fearful humans as they watched a burning log.

People always try to explain the things they don't understand. At first, this was done with stories and guesses. Later, it was done through scientific study. The Greeks were trying to explain fire 2,500 years ago. They divided all matter into four elements—earth, air, water, and fire. For 2,000 years fire was thought to be a separate element.

Do you know what phlogiston is? If you had lived in the early 1700's, you would. By this time scientists had a new idea, or theory, about fire. They decided that some materials contained an element, phlogiston, that would burn. Unless phlogiston was present in a material, they thought, the material wouldn't burn. Wood and paper contained phlogiston, so they would burn, but rocks had none, so they wouldn't burn. When something burned, according to this theory, phlogiston was released into the air and the object became lighter.

Antoine Lavoisier, a French chemist, was checking this theory in 1772. He weighed all elements (ash, soot, smoke) before and after burning something. He found that burning caused an object to become heavier! Rather than releasing something into the air, he reasoned, the burning material was absorbing some element in the air.

Joseph Priestley, an English chemist, discovered a new gas in 1774. Lavoisier decided that this gas, which he named oxygen, was present in air and was being absorbed by burning material when something was burned.

At last scientists were on the right track. But to understand exactly how combustion, or burning, takes place, we have to go back and take a look at another old theory. The ancient Greeks had first proposed that everything was made up of particles. They called these atoms. In 1808 another English chemist, John Dalton, added to this theory. He believed that each kind of atom had a different size and weight. Atoms, he thought, could be combined in different ways to form molecules, tiny particles of matter. For example, one atom of oxygen and two atoms of hydrogen form water molecules. He theorized that atoms were held together in the molecules by chemical bonds.

Today we know that everything on earth is made up of atoms formed into molecules. Matter is found in three states—solid, liquid, or gas. In a solid, the molecules are close together and there is little movement. In a liquid, molecules are farther apart and there is more movement. In a gas, molecules are still farther apart and move very rapidly.

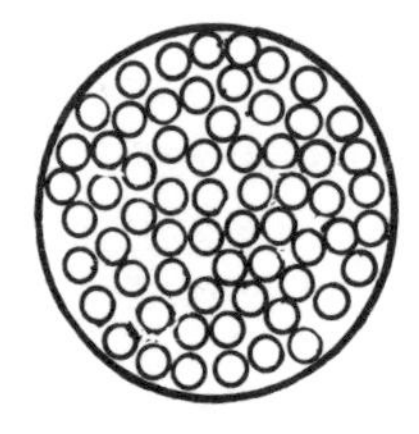
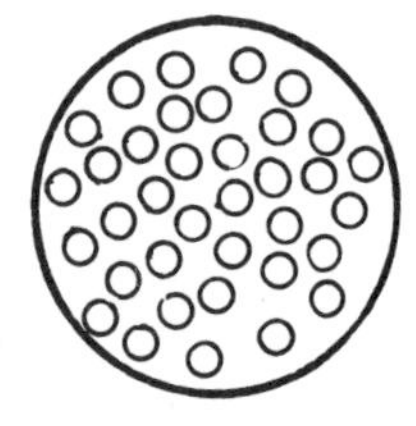
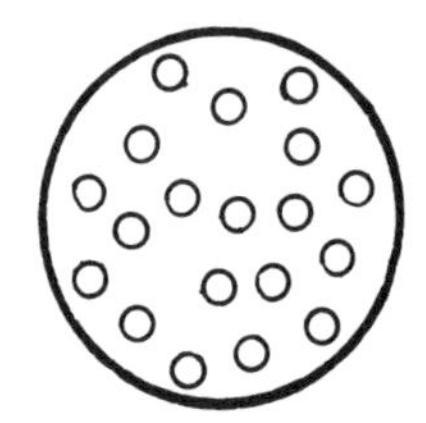

Molecules in three states—solid, liquid, gas

Heat causes molecules to separate and move about more rapidly. When they become hot, molecules dash around so wildly that they break their chemical bonds and split into separate elements, or atoms. Oxygen combines with these elements being released as gas so rapidly that it causes small explosions. Some energy is released in the form of the light we see. Energy in the form of heat is also released. Fire is the heat and light given off when atoms of oxygen combine rapidly with other atoms.

OXYGEN + FUEL + HEAT = FIRE

What three things are necessary to start a fire? First, you will probably say, something to burn, and you will be right. Fuel, or something to burn, can be in any one of three states—solid, liquid, or gas. Examples of solid fuel are wood and coal. A liquid fuel could be gasoline or oil. A gas fuel could be hydrogen or natural gas. Some fuels, such as gunpowder and dynamite, burn extremely rapidly. They produce such volumes of gas that they take up much more space than a solid. This violent expansion of gas is an explosion.

You cannot start a fire until you have oxygen to combine with other elements. Air, which is one fifth oxygen, usually furnishes the oxygen. Fuels burn much more rapidly in pure oxygen, and some things will burn only when pure oxygen is present.

One other thing is necessary. There must be enough heat to raise the temperature of the fuel to the point where it will start burning. The kindling temperature, as this point is called, is the lowest temperature at which a substance takes fire and continues to burn.

Fuels have different kindling temperatures. Phosphorus will ignite from the warmth of your hand. Paper, dry wood, cotton, and wool ignite at a higher temperature. Some fuels, such as coke or hard coal, have

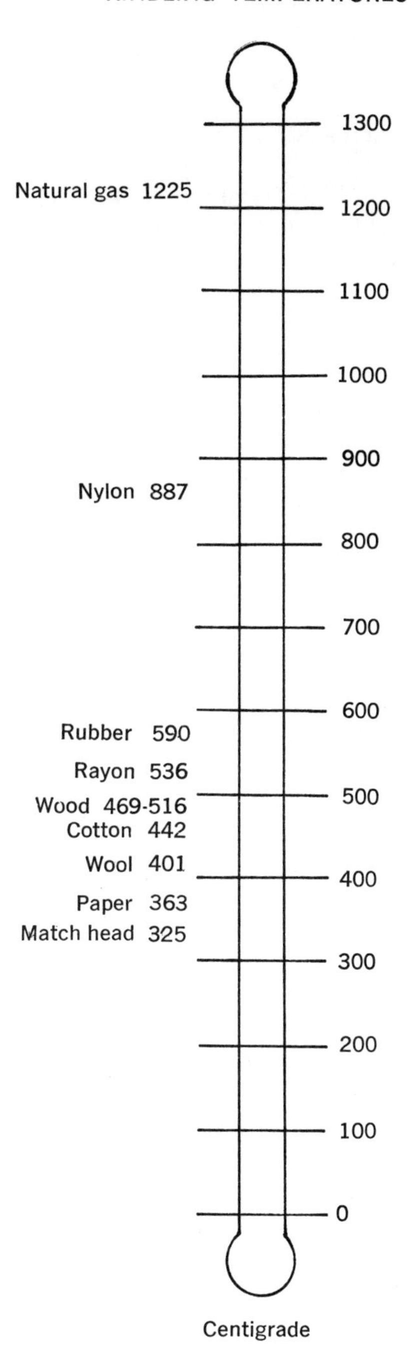

such a high kindling temperature that they do not ignite unless placed in a fire.

The size of the piece of fuel also affects the kindling temperature. It takes more heat to ignite a large piece of fuel. A match will set a twig on fire, but another fire is required to start a large log burning.

FIRE TRIANGLE

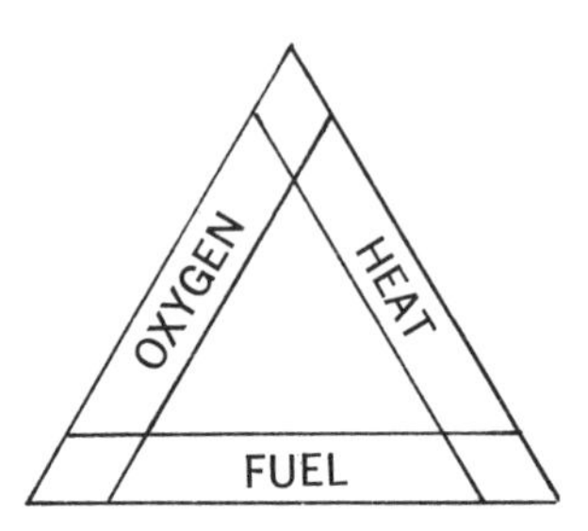

The fire triangle—oxygen + fuel + heat = fire

Friction, or the rubbing of two things together, can produce enough heat to provide a kindling temperature. This method was often used by American Indians and other primitive people. We start most of our fires today by friction. Match heads are coated with a chemical that has a low kindling temperature. When it is rubbed against a rough surface, friction produces enough heat to start the chemicals on the match head burning. Then the heat of one fire can be used to start another fire.

Oxidation is the process by which oxygen unites with another element. Heat, as we have seen, causes this to occur very rapidly. The result is fire. Without heat at the start, oxidation takes place at a slower rate. But heat is always produced and, if it cannot escape, builds up until it reaches the kindling temperature. Then the material bursts into flames. This process is called spontaneous combustion.

Thousands of fires start every year from spontaneous combustion. Oily rags stored in a confined space often catch fire. Oxidation also takes place as hay is being cured. If hay is stored in a barn before it is completely dry, the result may be a barn fire.

TYPES OF FIRES—HOW THEY ARE PUT OUT

Do you remember the fire triangle? The way to put a fire out is just the opposite of starting one. You must remove one side of the fire triangle. In some fires the best way to put a fire out is to remove the fuel. This is the system used in fighting forest or large grass fires. Bulldozers and men go in to remove trees, brush, and dry grass in the path of the fire. When the fire reaches the cleared place, there is nothing to burn, so the fire goes out.

This system is also used when fighting a conflagration in a city. Conflagrations, huge fires covering large areas, sometimes occur after an earthquake or a bombing raid during a war. They can also occur when fire breaks out among crowded wooden buildings, especially if there is a high wind. If the heat is intense, the burning buildings on one side of the street will ignite the buildings across the street. Sometimes buildings must be dynamited and burnable material bulldozed out of the path of the fire before it can be stopped.

A fire cannot burn without oxygen. If the supply of oxygen can be cut off, the fire dies out. This is the method used when a blanket is thrown around someone whose clothes have caught fire. The blanket cuts off the supply of oxygen. The victim can also roll on the ground and do the same thing.

Grease being used in cooking sometimes catches fire. If a tight cover is put on the pan, the fire goes out. But in most fires there is no cover large enough to cut off the oxygen. It must be shut out through other methods.

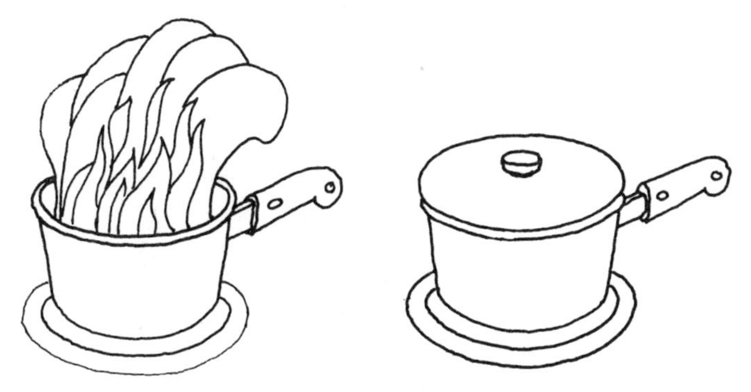

Fire in oil or grease can be put out by shutting off oxygen supply

Fire extinguishers contain chemicals that are used as a foam or a powder to smother a fire. Dry powder is used on electrical and motor fires and on burning metals such as magnesium. It is dangerous to use any type of extinguisher that contains water on these types of fires. Electricity can travel up a stream of water and electrocute the fire fighter. Water causes magnesium to burn more fiercely.

Foam is used on burning liquids such as gasoline and oil. Since these liquids are lighter than water, they float on water and spread the fire. Chemical foam will smother many types of fire and is used in most home and office fire extinguishers.

There is one more way to put out a fire. This is to cool the fuel until it is below its kindling temperature. Water is usually used for this purpose. Water sprayed on many burning materials such as wood, paper, or cloth will cool the temperature enough so they will not burn. Sometimes chemicals are added to water to increase its cooling ability. Water mixed with chemicals is dropped from bombers to cool a forest fire down enough for men to get close to fight it.

In all fire-fighting methods, firemen try to do one of these three things —remove the fuel, cut off oxygen, or cool the fuel. Some agents act upon a fire in two ways. They coat the fuel to help cut off oxygen and, at the same time, lower the temperature. As you read this book or watch firemen in action, ask yourself which method is being used. Which side of the fire triangle are the firemen trying to remove?

Not all fires can be fought the same way. Different locations and materials present different problems. The same fire-fighting methods cannot be used on a city fire and a forest fire. Let's look at different types of fires and how they are fought.

3
CITY FIRES

PROBLEMS IN FIGHTING CITY FIRES

In New York City no one glances up when a jet passes overhead. Airplanes are taking off and landing at nearby airports twenty-four hours a day. But on the morning of July 28, 1945, the shoppers on Fifth Avenue were startled by the roar of a low-flying plane. Since all pilots have instructions to maintain an altitude of more than 2,000 feet over New York's skyscrapers, this was unusual. Looking up, many persons could see an Army B-25 bomber flying much too low over Manhattan.

The Empire State Building towered in the flight path of this aircraft. As the shocked pedestrians watched, they saw a flash and were shaken by a powerful explosion. The airplane had plowed into that skyscraper at the seventy-eighth and seventy-ninth floors. The bomber tore an 18 × 20 foot hole in the building and landed inside on the seventy-eighth floor. Flaming gasoline was sprayed over the outside of the building for five stories above and below the crash site. One engine tore out an elevator shaft, dropping one elevator seventy-eight floors. Burning debris fell down the elevator shaft, starting another fire on the ground floor.

But the damage wasn't restricted to the Empire State Building. Part of the burning plane careened through the building and out of a window on the other side. It landed on a penthouse on a twelve-story building. That too was set afire.

Four alarms were sent in to the New York City Fire Department within eight minutes. Twenty-three fire companies responded to fight the three separate fires started by the crash.

City fires like this one in the Empire State Building present special problems and require specialized equipment. The upper floors in skyscrap-

A view of one of the floors of the Empire State Building after the crash of the B-25 bomber
UNITED PRESS INTERNATIONAL

ers are far beyond the reach of streams of water from the ground or from water towers. The tallest aerial ladders will only reach to the tenth floor. In older buildings the only way to get water to the upper floors is to drag the heavy hoses up many flights of stairs. Elevators cannot be used. The buttons on automatic elevators are usually heat activated. The heat causes the elevator to stop at the floor where the fire is located. The doors open and will not close again.

Terrible fires in older high-rise buildings have resulted in strict fire laws in most cities. Modern skyscrapers are engineered to be fire-resistant. They are built of masonry or concrete reinforced with steel. All steel beams are enclosed in concrete to prevent buckling during a fire. The floors, roofs, and interior walls are made of material that will not burn or fail under intense heat. Standpipes, large water pipes to carry water to the top floors, are built into the building. You have probably seen the two standpipe outlets on the front of high-rise buildings.

Fortunately for the thousands of busy workers that morning, the Empire State Building had excellent built-in fire-fighting facilities that were well maintained. Nine standpipes had been built into the skyscraper. Two of them went to the eighty-third floor. A pumper was quickly hooked up to the standpipes on the ground floor and pumped water to the upper floors from a fire hydrant. There were interior water outlets on every floor. Fire hoses were always kept hooked to these. The Empire State Building had six miles of hose hooked up and ready to fight a fire.

The building had huge water tanks and pumps located in the basement and on different floors. Although the flaming gasoline had started an intense fire, firemen were able to get enough water to it to prevent a disaster.

The loss from this fire was only $500,000, and only 20 percent of the loss was caused by the fire. Most of the damage was done to the building by the crash. The loss of life was kept low. The three-man crew of the

Tower of Empire State Building enveloped in smoke and flame after the crash of the B-25 bomber

UNITED PRESS INTERNATIONAL

bomber died instantly. Eleven women clerks, working on the seventy-ninth floor where the airplane hit, were killed by the flaming gasoline. Although this was a working day and thousands of people were in the building, no other lives were lost.

Montreal firemen use aerial ladders to rescue 250 office workers trapped after an explosion and fire on the second floor of the Canadian Liquid Air Building

UNITED PRESS INTERNATIONAL

Heat and smoke build up to temperatures of 720° to 810°C. (1200°F. to 1500°F.) when there is a fire inside a modern skyscraper. The windows are not meant to be opened. Often the pressure of expanding air and gas builds up until the windows are blown out. Firemen never try to break these heavy windows with their axes. The glass, which is about 3.75 centimeters (1½ in.) thick, picks up momentum as it falls from high buildings. It often travels the distance of a block or more and hits with such force that people have been decapitated by falling glass.

Let's look at the special equipment used by city fire departments to fight their fires.

SPECIAL EQUIPMENT FOR FIGHTING CITY FIRES

Aerial Ladders

Sirens wail as fire engines race through the dirty narrow city streets. The glare in the night sky becomes brighter and brighter. The tillerman at his post on the end of the swaying aerial-ladder truck steers the rear wheels around the last sharp corner. He feels like the end person in a crack-the-whip game.

Aerial ladder being used to fight a five-alarm fire in Boston
UNITED PRESS INTERNATIONAL

As the driver pulls up in front of the burning tenement, the tillerman's practiced eyes search the upper-floor windows. Yes, there are people trapped up there on the fourth floor. He can see their forms outlined against the glow inside. As he jumps to the ground, the tillerman snatches the big steering wheel off so the ladder can be raised. In a few seconds the jacks have been lowered under the truck to steady and support the weight of the huge ladder. The driver pulls the lever that starts the ladder on its long rise.

Firemen start climbing the ladder while it is going up to be ready to help the terrified people at the windows. The driver stays with the truck and operates the lever to the turntable on which the foot of the ladder rests. He must keep his eyes on the firemen on the ladder and on the burning building at all times. He turns and moves the huge ladder. If people are waiting at other windows, the operator moves the ladder to another place while the first people are climbing down.

Some aerial ladders are as long as 44 meters (144 ft.). With the aid of these, firemen can rescue people in buildings up to ten stories high. Life nets can be used safely only for jumps from the second or third floors. Beyond this height, a body falls with too much force for the firemen to keep it from hitting the ground.

Aerial-ladder trucks are an important part of all big-city fire department equipment. Since the ladders are hydraulically operated, one person can position and move them. It takes four to six firemen to raise a 12 meter (40 ft.) ladder by hand. A nozzle mounted at the top of the ladder, with a fire hose running up to it, is used to get water to the upper floors. Aerial ladders are also used to spray water on lower roofs to keep them from catching fire from a burning high rise.

Snorkel

The gooseneck crane, commonly called the snorkel, is another piece of aerial fire-fighting equipment. The snorkel does not have a ladder. It consists of two 12 meter (40 ft.) hinged booms resting on a hydraulically operated turntable. Water pipes and nozzles are built into the snorkel. Firemen ride in a metal basket that can be raised, lowered, or moved as they work. This makes an aerial platform from which to fight a fire. The snorkel is also an important piece of rescue equipment as it can carry several people at one time. If necessary, stretchers are laid across the basket to bring the injured to the ground.

Chicago firemen using snorkel to fight a fire

UNITED PRESS INTERNATIONAL

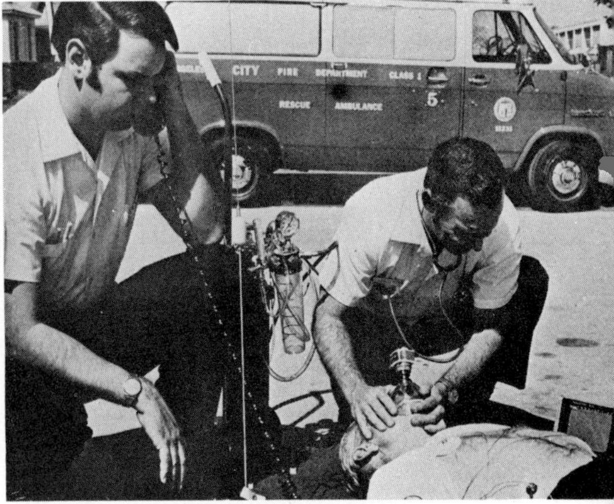

The rescue squad is part of most city fire departments

LOS ANGELES FIRE DEPARTMENT

An effective communications system is vital to a city fire department

LOS ANGELES FIRE DEPARTMENT

Water Towers

Many large-city fire departments still have and use the older water towers when fighting fires in tall buildings. The water tower is carried on a truck and raised at the fire in the same manner as an aerial ladder. It consists of a heavy-stream nozzle mounted on top of a tower. The operator on the truck can raise, lower, or turn the nozzle to direct the stream of water. Water towers are no longer manufactured and are being replaced by aerial ladders and snorkels.

Superpumper

Fighting a large fire can deplete a city's water supply. During disasters such as air raids or earthquakes, there is always the chance of the water mains being damaged.

Superpumper is the answer to a city fire department's need for a huge quantity of water. It is an enormous pumper designed to pump water from a river, harbor, pond, or lake and deliver it to the fire under high pressure. Superpumper could be called a fireboat on wheels.

Supertender travels with Superpumper carrying a half mile of special high-pressure hose. One end of this hose can be dropped into a harbor so water can be pumped to a fire 600 meters (2,000 ft.) away. Satellite trucks, with water cannons mounted on their cabs, also accompany Superpumper. The cab of Supertender, with a water cannon mounted on top, is disconnected so it can be moved closer to fight the fire. Supertender and the Satellite trucks draw their water from Superpumper. Together they can throw 32 metric tons of water a minute.

The water cannons mounted on the cabs of all of these trucks can be turned in any direction. They shoot powerful streams of water 150 meters (500 ft.). The water hits a burning building with enough force to demolish a rickety wooden one. If directed inside, the water will knock down flimsy interior partitions and move burning material around so that the fire can be extinguished.

Superpumpers have been built only for a few large city fire departments. Their cost prevents any but the largest cities from ordering them.

Rescue Units

Most metropolitan fire departments also have an especially trained and equipped rescue squad that responds to any type of an emergency call. Its vehicles can double as ambulances.

Rescue squad firemen are trained in advanced first aid. Since they can respond quickly, they are often called when a medical emergency occurs. They use resuscitators, if necessary, give first aid, and take the victim to the hospital. Rescue units carry specialized equipment to help in all types of rescue work. They have gas masks to be used to enter a smoke- or gas-filled building. They carry inhalators, as well as resuscitators, to help revive drowning victims or anyone overcome from gas fumes or smoke. Portable floodlights and spotlights are used at night. Metal-cutting tools are used to help free people trapped in cars after an accident. Portable electric saws are carried to cut through wood, and an electric generator to provide electricity. Two-way radios or telephones help the rescue squad members keep in touch with others and with the firehouse.

Communications System

A city fire department cannot be effective without a good communications system. Fire stations are spread all over the city. Each one covers a certain area. When a fire alarm is turned in, the call goes to the central alarm station. There the call, time, and call box number are recorded on a permanent tape. The alarm is then relayed to the fire station nearest the fire. Police units are also notified to handle the traffic. Other fire stations are alerted so that they will cover if another fire should break out in that area.

All units of the fire department are equipped with radiotelephones for communication with each other and with headquarters. Fire engines responding to a fire can be rerouted or told to return to the firehouse if necessary. While fighting a fire, the fire chief uses portable two-way radiotelephones to keep in touch with headquarters. He also uses an electric megaphone to call directions to the firemen over the roar and crackle of the flames.

CONFLAGRATIONS

A fire that extends over a large area and involves many buildings is called a conflagration. This type of fire often occurs during wartime bombing raids. The central zones of many European and Japanese cities were destroyed by conflagrations following air raids during World War II. They can also follow natural disasters such as earthquakes.

There are several conditions that can contribute to a conflagration. One of these is wooden buildings built closely together—especially buildings with wooden roofs. A high wind or a large amount of flammable material stored near other buildings is another factor. A delay in discovering the fire or sending in an alarm, or an inadequate supply of water can cause a small fire to become a big one. Not all of these conditions have to be present at one time for a conflagration. But when several of them exist, there is danger of a fire's getting out of control.

During a conflagration, the intense heat of the burning buildings sets others on fire—even those across the street. Often the only way to stop a conflagration is to dynamite buildings and bulldoze the debris out of the path of the fire. The two great city fires described in Chapter 9 are examples of conflagrations.

4
FIRES IN TOWN AND COUNTRY

SMALL-TOWN FIRE DEPARTMENTS

Toot! Toot! Toot! . . . Toot! . . . Toot! Toot! The mechanic on the creeper under the Mack truck listens intently. There it is again. Toot! Toot! Toot! . . . Toot! . . . Toot! Toot! His lips move as he counts to himself.

"Three twelve!" yells Tom, the mechanic, pushing himself out from under the truck. "North and Main." Tom's behavior doesn't surprise his boss. He knows Tom is a volunteer fireman. Tom only takes time to wipe the grease from his hands and shed his greasy overalls before he jumps into his car. He can hear the sirens going down Main Street as he pulls out of the garage.

A woman is on the corner of North and Main waving a dish towel as the fire chief, followed by his two fire engines, pulls up.

"Over there!" she yells, motioning. "We were just trying to burn the dry grass, but the shed caught fire."

The shed is burning hotly. So is the dry grass in a wide area around it. The fire chief shakes his head. Another homeowner who doesn't realize how fast a grass fire can spread. The firemen will be lucky if they can save the house.

The nearest fire hydrant is a half mile away. But one of the pumpers contains a big tank of water. The other pumper pulls up to the ditch that runs in back of North Street. Rural firemen know the location of every body of water.

The volunteers are arriving now. They grab their fire hats and coats as they race to help. The fire chief yells instructions to them. These men have been trained in fire-fighting methods at weekly meetings. They know what to do. They get the suction hose hooked to the pumper and drop it into the ditch. Then they lay a line of canvas hose from the pumper to the fire.

The house could go any moment. The wall nearest the fire is already smoking and the paint is blistering. Several pieces of debris from the blazing shed are smoldering on the roof. But the firemen are lucky. They start pumping water on the house before it can burst into flames.

Within an hour the fire is out. Tom and the other volunteers roll up the hose and get the fire engines ready to go back to the firehouse. Then they leave for their usual jobs. The fire chief and the two paid employees of the fire department stay behind. They see that everything is cleaned up and there is no chance of the fire's breaking out again. The fire chief also gives the householder a little lecture on fire safety.

This small fire department is much like those located all over the United States in rural or urban areas. People and buildings are spread out here. There are fewer people paying taxes to support a fire department. A few professionals, the fire chief and a few others, are employed full time. But most urban fire departments depend on unpaid volunteers for most of their fire-fighting needs. Volunteers do this because they enjoy the excitement of fighting fires and the good feeling of helping others. They donate their time and pay their own expenses when they attend training sessions and go to fires.

When the fire chief sees that his company can't handle a fire, he puts out a call for help. Nearby rural fire departments respond with men and trucks. Sometimes they cover for the busy fire department in case another fire should break out. State or county fire crews often bring fire trucks if needed.

SPECIAL EQUIPMENT

Fire Engines

A small-town fire department probably has one firehouse with one or two fire engines. Larger towns, of course, have more. If a fire department has only one fire engine, it is probably a combination pumper and hose car. This truck also carries ladders, axes, extinguishers, and other fire-fighting equipment. If there are buildings in town higher than two stories, the fire department may have a combination pumper and aerial-ladder truck. This aerial ladder is much shorter than those used in cities.

Fire hydrants are few and far between in rural areas. Fire engines equipped with water tanks are part of many rural fire departments. Most of these are one-ton pickup trucks operated by one man. They usually have a 120-imperial gallon tank of water, a centrifugal pump, and a small diameter hose. They are equipped with powerful motors and special tires for rough ground. They can get to a fire far from roads where a larger truck would get into trouble. Sometimes they are supplied with refills of water from a larger truck parked farther from the fire.

The "brush breaker," another fire engine that carries its own water, is

Pack pump being used to put out fire in rural area
U.S. FOREST SERVICE

equipped with a tank of up to 800 imperial gallons. It is a high-clearance four-wheel drive truck built to travel over rough ground. If the water tank is large, it will be equipped with dual rear wheels to support the weight of the water. A steel shield is fastened in front of the wheels to break down brush and small trees so the truck can roll over them. The brush breaker can direct a strong stream of water to the base of the fire to cool it down. Then men with hand tools can move in to finish it off.

Other Equipment

Fires in rural areas often start in dry grass or weeds. In the early stages these fires can be fought with brooms, wet sacks, or anything with which to beat out the flames. Shovels are used to throw dirt on the fire and remove combustible material from its path.

Often rural fire departments have several "pack pumps" to be carried on men's backs. These 4-imperial gallon tanks each have a pump to shoot out a steady stream of water.

If there is a pond or stream near the fire, but the ground is too rough for a vehicle, a portable pump may be brought in. In this case the pump and fire hose usually have to be carried in by hand. One end of the suction hose is dropped into the pond and hooked to the pump. The fire hose is laid from the pump to the fire. The portable pump produces a strong enough stream of water to put out most fires or to cool them down.

If a grass or brush fire gets out of control, a call may be sent to a state or county fire-fighting crew to send out a bulldozer. This is carried on a flatbed truck to the fire site. Bulldozers are used to clear a firebreak. They remove the brush and grass so the fire will die out when it gets to the bare ground.

CONFLAGRATIONS IN RURAL AREAS

Brentwood and Bel Air are rural districts of Los Angeles. Here expensive houses dot the hills. Homes are often built on the sides of canyons. When this is done, one side of the house is at street level while the other is supported by stilts. Native shrubs are allowed to grow in the open place under the house to give it a woodsy look. Each householder owns several wooded acres that are left as natural as possible. Winding roads and dead-end streets lead to these picturesque homes with their heavy wooden shake roofs. It is a beautiful, woodsy, expensive, and dangerous place in which to live.

The Brentwood fire. A house stood here!
LOS ANGELES FIRE DEPARTMENT

The chaparral that grows on these hills is a fire hazard. Chaparral is the growth of heat- and drought-resistant small trees and brush that covers many of the western hills. Chaparral has an extremely high oil content. It is probably the most combustible brush in North America. Before the area was heavily populated, chaparral burned over every few years. The larger trees usually survived, while the brush and dry grass were cleared out to grow again when it rained. But fires are not allowed to burn unchecked there now. The chaparral grows dense and tall.

Southern California has several conditions besides the chaparral that make it a hazardous fire area. It has only about 25 centimeters (10 in.) of rainfall a year—usually in the fall and the winter. During the dry summer, six months may go by without rain. The heat may reach 43° C. (110 F.). There is, at times, another dangerous condition there. Occasionally a strong northeast wind blows down the mountain passes and out to sea. This wind, known as the Santa Ana, or "devil wind," usually blows for several days at a time. When it occurs during dry hot weather, conditions are ripe for a conflagration.

Conditions were right for a big fire on November 5, 1961. It had been the driest summer and fall in Los Angeles history. The humidity was down to 9 percent and the Santa Ana was blowing at 46 kilometers (29 mi.) an hour.

The spreading city of Los Angeles covers miles of metropolitan and rural areas. It is organized into one fire department with hundreds of substations. All the firemen who worked for the city that day were uneasy. They knew that fire conditions were unfavorable. The fire chief declared an extra hazard day and moved many of his units into the hilly suburban areas to reinforce the regular crews stationed there. They were expecting trouble and didn't have long to wait.

At 10:10 A.M. the first alarm came in from a one-acre brush fire at a garbage dump. The firemen who responded never had a chance to contain that fire. Within fifteen minutes five more calls had gone out and thirteen fire engines had responded. By then the fire was racing through the chaparral, blown by winds of up to 80 kilometers (50 mi.) an hour. Clouds of burning embers were swirled about by the winds and blown onto dry wooden roofs some distance away.

Fires of this type leapfrog from place to place. The dry shake roofs burned almost explosively, sending more clouds of embers skyward. Even if a resident had cleared a firebreak around his home, the heat of the burning chaparral and the flying embers ignited the house. The fire burned up under homes built on steep slopes so that they burned from below as well as from above.

A borate bomber dumping fire retardant on a brush fire in Woolsey Canyon, Calif.

UNITED PRESS INTERNATIONAL

A helicopter being used by the Los Angeles Fire Department to fight a brush fire

LOS ANGELES FIRE DEPARTMENT

So many calls came into the Los Angeles Fire Department that there was no way they could all be answered. By 10:35 A.M. fifty-nine engines were at the scene of the fire. By 12:30 P.M. ninety-six fire engines had been sent out. There was no doubt about it by that time, Southern California was faced with a dangerous conflagration.

A call for mutual aid was sent out. County fire engines and those from surrounding towns began arriving. The state Forest Service was asked to help. By that afternoon 154 fire engines with their crews were battling the blaze. Fifty-four other fire vehicles were also on the fire line. Fourteen air tankers were at work dropping a mixture of water and chemicals on the fire. The battle was on.

The pilots of the air tankers provided other valuable help. They used their radios to send back information about new fires as they broke out. At 4:00 P.M. one pilot reported a small fire in Mandeville Canyon. So far this area of expensive homes had been spared by the fire. When this information came in, fifty-five engines were sent to Mandeville Canyon. The firemen caught that fire in its early stages and stopped it cold.

By four o'clock the next morning, the fire was under control. But it had been the worst fire in California since the great earthquake and fire in San Francisco in 1906. Property damage was estimated at 25 million dollars with over 500 homes destroyed. This makes it the fifth most costly fire ever to occur in the United States. Fortunately no lives were lost.

Because of its dry summers, California has suffered more than other states from these rural conflagrations. Most have occurred in Southern California, but some have occurred in other locations. On September 17, 1923, a brush fire broke out near Berkeley, California, across the bay from San Francisco. This fire destroyed 640 buildings on the outskirts of that town.

There have been rural conflagrations in the East. On a windy day in April, 1941, a big grass fire broke out near Marshfield, Massachusetts. It burned 450 buildings and couldn't be put out until it reached the shores of the Atlantic Ocean.

5
FOREST FIRES

GUARDING AGAINST FOREST FIRES

When Beth crawled out of her cot in the morning, the sun was just lighting the valley below. Beth scanned the hills and the valley, looking for any telltale column of smoke. She did this every morning before she washed and dressed.

This was Beth's second summer of watching for fires from a small lookout perched on top of Eagle Rock. Beth intended to work full time for the Forest Service when she was through college and this was good training. Besides, she loved the work. When school was over in June, the Forest Service truck had hauled Beth and all her necessities up the steep rough road to the lookout on Eagle Rock. Every other week the truck would groan back up the hill, bringing fresh food, water, and mail to Beth. The rest of the time she was alone except for the radiotelephone. She talked with the fire control base and the people at other lookouts every day. Beth would stay there until the fire season was over. Then she would go back to school.

Lookouts are manned only during the summer fire season, since forest fires only occur under certain conditions. Wet and snow-covered forests don't burn. But when the weather is hot and the humidity is low, conifers, with their high resin content, are extremely flammable. The lookouts, as well as the state and federal Forest Service fire-fighting stations, are manned during the summer.

Beth was worried as she picked up her binoculars. A strong wind was blowing that morning. There had been no rain for over a month. Any spark could set off a major forest fire. Beth had been awakened by thunder rumbling during the night. Occasionally the lookout had been lighted by a flash of lightning. Beth knew that most forest fires there in the West were started by lightning. In summer, lightning often occurs in the West with little or no rain. In the East, lightning is usually followed by a heavy downpour.

Now Beth focused her binoculars on the spot in the trees where a campground was hidden at Bear Lake. Forest fires were often started there by campers, fishermen, and hikers vacationing in the summer. Careless smokers and children playing with matches were other big dangers. But the smoke coming from the vicinity of Bear Lake that morning seemed to be from cooking fires.

Now Beth studied the crests of the hills. She moved the glasses slowly along one crest after another, looking for a telltale plume of smoke. The wind made smoke detection difficult. Then a spot near the crest of Bald Mountain drew her attention. She watched that spot intently for almost five minutes. Yes, there was no doubt about it. There was smoke coming from a ravine near the crest of Bald Mountain.

Beth turned to the azimuth indicator, a device to help her obtain the compass direction of the smoke. She carefully lined the indicator up with the smoke. Now she had a fix. If another lookout reported the same smoke, the fire control base could pinpoint its location.

Beth picked up the radiotelephone.

"Eagle Rock to headquarters. Eagle Rock to headquarters."

"Come in, Eagle Rock."

"I have a fix on smoke from a ravine near the crest of Bald Mountain."

"Good! Fire Mountain just called in to report that smoke. Give me your reading."

FIGHTING FOREST FIRES

While Beth was eating her cornflakes, she heard a small airplane overhead. It was a spotter sent out by the Forest Service to survey the situation. The slopes of Bald Mountain were so steep, fire fighters would have trouble getting in. There were no roads and it would take hours, or even days, for fire fighters to hike in.

Beth watched as the spotter plane circled the fire several times and then left. There was more smoke now. The fire was evidently working its way up the ravine. Beth wiped the perspiration from her face. It was 10:00 A.M. and already warm.

In about an hour Beth heard the sound she had been expecting. A DC-3 was flying toward the fire. Beth knew what that meant. The Forest Service was sending in smoke jumpers.

Smoke jumpers prepare to leave their plane
U.S. FOREST SERVICE

Beth had visited the smoke-jumping base about fifty miles away. She had watched the smoke jumpers' rugged training and had met many of the men. They had even let her fly with them on one fire call. She could picture the interior of the plane with all the smoke jumpers ready to jump.

Smoke jumpers are usually college students with fire-fighting experience. They must pass a rigid physical examination before they are accepted into the program. Smoke jumpers are not big men. The maximum weight for smoke jumpers is 81 kilograms (180 lbs.). But they must be in excellent physical condition.

Beth watched the plane through her binoculars. First the plane passed over the fire. The jumpmaster at the open door threw out some crepe paper streamers and watched them float down. He was estimating what effect the wind would have on the parachutes. He was also looking for an opening in the trees where the men could land.

A smoke jumper descending toward a forest fire. Steering slots and lobes allow him to direct his fall by pulling dark guidelines
U.S. FOREST SERVICE

Now the plane circled and came back for another pass at about 450 meters (1,500 ft.). Beth could picture the smoke jumpers in their heavy canvas coveralls, helmets, and face masks. Their large parachutes would be strapped to their backs with the emergency chute in front. The huge pockets in the legs of their jump suits would be bulging with equipment to fight the fire. Since they would be out several days, other supplies would be dropped to them. She knew the first smoke jumper would be standing in the doorway ready to dive out. His chute would be hooked to the static line so that it would open automatically as he left the plane. Another jumper would be right in back of him ready to follow. They would be waiting for the jump-master to give the word.

There—Beth saw the first parachute appear. It was followed almost immediately by another. Only two men jumped on the first pass. The jumpers were trying to land in a small meadow below the fire. The plane had to be in the right place when they jumped. Beth watched through her binoculars. She could see the men working their shroud lines to try to hit

Smoke jumpers landing at a fire. Notice the chutes hung up in trees

U.S. FOREST SERVICE

the small landing site and avoid trees. One smoke jumper hit the meadow and rolled as he landed. He quickly jumped up and spilled the air out of his chute so that it wouldn't drag him along on the ground. The parachute of the second smoke jumper was caught by the wind. It got hung up in a pine tree and left him dangling 7.5 meters (25 ft.) from the ground. He would use a rope to rappel himself to the ground.

By now the plane had dropped two more smoke jumpers and was circling to drop two more. Beth knew the smoke jumpers would shed their heavy coveralls as soon as they were on the ground. They would leave these and their parachutes to be picked up later. Now they would take their fire shovels and try to put out the fire.

Beth watched the fire all day. The smoke jumpers didn't seem able to control it. The brisk wind was carrying it up the slope. Later in the day a cargo plane flew in low to drop more equipment attached to cargo chutes. The smoke jumpers must have radioed out for more supplies.

By the next morning Beth could tell that the fire had spread. The wind had shifted and the fire was sweeping along the ridge. Beth saw the small spotter plane flying around again, checking the situation. Soon she heard the motor of a heavier plane. The Forest Service was sending in some borate bombers. Beth watched as the air tanker dropped its load of fire retardant at the head of the fire. Red dye in the mixture of chemicals and water made it easy to see where the mixture had landed. The air tankers made many passes that morning, trying to cool down the fire so that the smoke jumpers could move in and finish it off. More smoke jumpers and equipment were sent in later that day. In the afternoon Beth saw a helicopter laying fire hose for the fire fighters. They must have dropped a pump so the men could pump water from Bear Creek.

By the next day the fire was under control. But it had swept across the ridge and to within a few miles of the road leading up to Eagle Rock. Beth watched as the Forest Service brought in fire fighters by truck to mop up the fire. It could be a day or two before the smoke jumpers could hike out carrying their parachutes, coveralls, and supplies. A helicopter would have to pick up the hose and pump.

METHODS OF FIGHTING FOREST FIRES

Most forest fires are put out by the ground, or hotshot, crews. They are the squads of Forest Service fire fighters who live and work at the fire stations during the summer. Since this work demands good physical condi-

Lingering forest fires are put out with shovels and Pulaskis
U.S. FOREST SERVICE

tion and endurance, only young people are employed. Many of them are students. Members of the hotshot crews must be at least eighteen years old. Women, as well as men, are accepted if they can do the work.

Men of the Hopi, Zuni, Apache, and other Southwestern Indian tribes make a business of fighting fires. Since they are accustomed to working in

The fire crew makes an X where it wants the equipment dropped

U.S. FOREST SERVICE

rugged country, they have become very skilled in this work. A Spanish-American team, Los Toros, from Mountainair, New Mexico, is also famous for its fire-fighting ability. The Forest Service often flies these teams in to fires when they need extra help.

Fires can be cooled down by drops of chemicals and water from air tankers, but it takes a ground crew to put it out. Most forest fires are put out with a shovel and a Pulaski—a combination curved hoe and ax. The Pulaski is used to clear out brush and small trees to make a firebreak. The shovel is used to throw dirt on the fire to smother it. After a little practice, a fire fighter can do this with amazing accuracy. Fire fighters can also use their shovels as emergency frying pans to cook bacon and eggs.

FROM THE AIR

Spotting

There aren't enough lookout towers to watch all the forests. In many locations a small spotter plane or a helicopter patrols daily during the fire season watching for telltale signs of smoke. Even when there is a lookout tower, spotter planes are also used. Often the observer in the lookout is too far away to tell much about the fire. A Forest Service employee rides along with the pilot to observe the fire and make notes. He notes the direction, the speed, and the intensity of the fire. He notes the kind of vegetation and the terrain where the fire is burning. He also checks to see how close the fire is to roads or trails leading into the area.

During the fire, the fire boss often flies over the site. From the air it is easier to get an idea of how the fight is going and where the men would do the most good. He looks for the best places for air tankers to drop fire-retardant mixtures or for smoke jumpers to land. Photographs are often taken to be studied later. An infrared unit is sometimes used. This unit can penetrate the smoke to photograph the hot areas.

Transporting Men and Equipment

How does the Forest Service get equipment and supplies in to fire fighters in rugged country without roads? The equipment is placed in padded bags to protect it and dropped with parachutes from low-flying planes. The parachutes are surplus or condemned personnel chutes. The Forest Service has worked out a color code for equipment drops. A different colored parachute is used for each kind of equipment so that fire fighters know what is coming down. When a fire crew calls for an air drop, it shows the pilot where to drop the equipment by marking out a big X on the ground.

An air tanker dropping borate slurry on a forest fire
U.S. FOREST SERVICE

Helicopters are used to lower equipment that cannot be dropped. As we have seen, they also lay fire hose. Sometimes helicopters are used to bring hot meals in to tired fire fighters who have been on a fire for several days.

Helicopters are used to take men in to fires in rough country when the use of parachutes is too hazardous. In some areas of the Southwest, jagged rocks, heavy brush, and high winds make it too dangerous for smoke jumpers. Often the country is too brushy or rough for a helicopter to land. In this case helijumpers are brought in. Fire fighters dressed in heavy protective coveralls and masks, much like those of the smoke jumpers, are brought in by helicopters. While the helicopter hovers from 1.5 to 3 meters (5 to 10 ft.) above the ground, the men jump. Their equipment is dropped from the same height.

Helicopters are also used for rescue work. When a fire fighter is injured, a call goes out for a helicopter fitted with stretchers. The helicopter can fly the injured person directly to the hospital or to a waiting ambulance. Helicopters have also been used to rescue fire fighters who have been trapped by a fire.

Chemical and Water Drops

Fire-retardant chemicals are mixed with the water to be dropped on forest fires. Borate or bentonite is usually used. These chemical mixtures penetrate heavy foliage and make it resist fire longer than water alone. Red dye is usually added to the mixture so that the pilot and other Forest Service officials can see where the load landed.

Most airplanes used as air tankers are old Navy torpedo bombers. The Forest Service installs on the undersides of these planes special tanks to hold liquids. When the pilot is over the target area, he pulls a lever that opens the tank doors for the mixture of water and chemicals to fall.

Bombing a fire can be dangerous work. Chemical and water drops are not effective unless the plane comes in at a low level. Fires are usually in mountainous country where there are strong crosswinds, updrafts, and downdrafts. A forest fire always produces a strong updraft. When the pilot releases 160 imperial gallons of fire retardant from the tanks all at once, the plane suddenly becomes lighter and lurches upward. It requires skillful flying to do this type of work. Most pilots have had close calls.

Helicopters are also used to make water drops on forest fires. They can be equipped with tanks and then are known as helitankers. If helitankers are not available; the helicopter may just carry bags of water each equipped with a short hose and nozzle.

Floatplanes are sometimes used to drop water. They have the advantage of being able to land on a lake to refill.

6
FIRES ON OR NEAR WATER

FIRES ABOARD SHIP

On September 5, 1934, the *Morro Castle* left Havana, Cuba, for New York with 300 passengers and 200 crew members aboard. This new luxury liner was said to be the safest ship afloat.

It was an enjoyable voyage until the evening of September 8. The night before the ship was to reach New York, things began to go wrong. This was the night of the big event—the captain's dinner. But the captain was seized with an unexplained illness just before dinner and died suddenly. A temporary captain was appointed, and the ship steamed ahead at full speed into a howling northeaster. The newly appointed captain did not reduce speed in spite of the high winds and rough seas.

At 2:30 A.M. some crew members smelled smoke. A fire had broken out in a locked locker in the smoking room. One of the *Morro Castle*'s new safety devices was a fire warning system. A red light flashed on when the heat in any room reached a certain temperature. But the smoking room was one of the few rooms where the fire warning system had not been installed. Valuable time was lost because no one could find the key to the locker. By the time the locker could be broken open, the fire was too large to be put out with a fire extinguisher. Whatever was burning in the locker was extremely flammable.

The crew ran for a fire hose. But when it was turned on, there was no water. Because a passenger had once slipped on a wet spot caused by a dripping fire hose, all hose connections had been capped. The fire made rapid headway while the crew struggled to get water into the fire hoses.

One of the safety features of the ship was the fire doors. They could be closed to prevent a fire from spreading. But the crew didn't know where they were located or how to operate them. The fire doors were never used.

Now the passengers began to smell smoke and turned out in their nightclothes to watch the battle. At first they laughed and joked. No one seemed to realize the seriousness of the situation.

The fire spread so rapidly that many people on the ship believed it must have been set. During the fire the ship continued traveling at full speed on its usual course. This, and the high winds of the storm, fanned the flames. Now the substitute captain decided to make for the New Jersey shore, which was only 16 or 19 kilometers (10 or 12 mi.) away. A captain usually heads for shore immediately if a fire should break out.

As the fire spread, no one told the frightened passengers what to do. No officer ever directed the fire-fighting efforts of the crew. No radio messages were sent out for help. Only a few lifeboats were ever lowered.

The burned-out hull of the Morro Castle *after being beached at Asbury Park, N.J.*

UNITED PRESS INTERNATIONAL

Those that were carried only a few people—mostly the crew. The chief engineer, too terrified to take command of the engine room, climbed into a lifeboat.

At last the crew got the hoses ready to go. But they were turned on all over the ship at the same time. Only a trickle of water came out. There wasn't enough pressure to supply all the hoses.

Twenty minutes after the fire broke out, another ship saw the fire and radioed the Coast Guard. But no message was sent out from the *Morro Castle.* It wasn't until 3:26 A.M. that the radio operator, George Rogers, sent out an S O S call giving the position of the ship. By that time, the radio room was on fire and the floor had buckled from the heat.

By now, most of the passengers had fled to the stern of the ship, where they were trapped by the fire. The wind blew the fire and smoke toward them. The captain and most of the crew, realizing it would be safer there, were on the prow. To save themselves from the fire, many passengers began jumping into the water in spite of the storm. The few lifeboats that were lowered never stopped to pick up the people struggling in the water. About 4:00 A.M. rescue boats began to arrive to pick up people in the water and those still trapped on the ship. Many were already dead.

After the *Morro Castle* fire, the radio operator went on a tour as the hero of the disaster. He claimed to have saved many lives by sending out the S O S call at 3:26 A.M. But there were too many things about the fire that could never be explained. The investigation board suspected, but were unable to prove, that George Rogers had something to do with the disaster. A few years later, in a separate incident, he was convicted of three murders. Although it was never proved, there was a suspicion that George Rogers caused the captain's death and the fire. His only motive seems to have been a desire to play the hero.

The fire in the *Morro Castle* is one example of why sailors fear the cry of "Fire!" above everything else. It is easy to see the problems of a fire on board a ship at sea. The fire is confined to an enclosed hull, where smoke and heat quickly build up. Diesel fuel and other flammable materials are carried on board ships. There is no way to escape from a fire at sea except into an open boat. The crew cannot put in a call for help from a well-equipped and trained fire department. Trained firemen keep their heads in the dangerous situations involved in fighting fires. But the crews of ships usually have none of this type of experience. Sometimes they become frightened and think only of saving themselves. For all of these reasons, fires on the water are extremely dangerous.

Today, the United States has strict laws requiring fire safety equipment and fire drills on all ships leaving United States ports. But these are not required by all countries. As the fire in the *Morro Castle* shows, safety features are of no value unless they are operational and are used.

WATERFRONT FIRES

The *Markay* was one of the largest petroleum tankers on the Pacific Coast. Early one Sunday morning in June, 1946, the *Markay* was lying at anchor next to an oil storage terminal in Los Angeles harbor. Although it was 2:00 A.M., the dock area was brightly lighted and workmen scurried around. The *Markay* was being filled with aviation fuel—gasoline and a butane blend. Almost 2.5 million imperial gallons of fuel had already been pumped aboard. The workers paid little attention to the fumes that hung over everything. They were used to loading flammable cargo. Two sides of the fire triangle were present here. There was fuel and oxygen, but there was no spark to set it off. The crewmen not on duty slumbered in their bunks in the *Markay*.

At 2:06 A.M., the *Markay* exploded with a roar that shook most of the Los Angeles area. Someone had carelessly supplied a spark to ignite the fuel. Probably a workman had tried to sneak what turned out to be his last cigarette. Flames shot hundreds of feet into the air. In a few minutes, the first explosion was followed by an even worse one. Twelve of the *Markay*'s twenty-one crewmen died in those first explosions.

The hull of the *Markay* was split from prow to stern. It looked as if a giant had brought a huge ax down on it. A third explosion quickly followed, sending up a huge mushroom-shaped cloud that hung over the ship. Pieces of metal and red-hot rivets were sent flying through the air. A 6 × 9 meter (20 × 30 ft.) piece of the tanker soared over all the oil tanks lining the docks and flattened a small building .4 kilometer (¼ mi.) away. In San Pedro, 8 kilometers (5 mi.) away, a woman was awakened by the blast. She was puzzled by a pattering sound outside and ran to the window. Her lawn and house were being pelted with red-hot rivets.

The *Markay* released thousands of gallons of blazing fuel. It floated on the water, and the tide quickly carried it under piers and warehouses. Within a few minutes, a mile of wharfage and several warehouses were in flames.

Waterfront fires often start with a bang like the one on the *Markay* and then burn fiercely. Many refineries, chemical plants, and other industries handling flammable materials are located on the waterfront. A variety of

Conventional fire-fighting equipment cannot reach fires on or over water

LOS ANGELES FIRE DEPARTMENT

flammable materials is also transported by ships which, of course, tie up to docks to load. In addition, the creosote used to prevent decay in wooden dock pilings is very flammable. There is another danger to dock fire fighters. Huge traveling cranes for loading cargo are located on the docks. If the dock supports are weakened by fire, the cranes topple over onto nearby structures and fire fighters. Waterfront fires are dangerous and difficult to put out.

Fireboat battling a dock fire. Side jets hold the vessel in place
LOS ANGELES FIRE DEPARTMENT

In the last few years a new fire hazard has been added to the waterfront. Thousands of small boats for recreation are used by inexperienced sailors. In small-craft harbors these boats are moored close together. Burning fuel from one boat quickly floats over the water to all boats and nearby structures.

How does a fire department fight one of these dangerous waterfront fires? Usual methods can only keep the fire from spreading to other buildings on land. Fighting fires on the waterfront requires different types of equipment and a different form of attack. How is it done?

FIREBOATS

When the fire alarm came in after the explosion on the *Markay,* the Los Angeles Fire Department sent out thirty pieces of equipment. But these firemen had no way of reaching the fire on or over the water. They worked trying to save oil storage tanks, warehouses, and wharves. A call was sent out for Fireboat 2 of the Los Angeles Fire Department.

The crew of Fireboat 2 already knew about the fire. Their berth was in the same area on the waterfront. The explosion had shaken the pilings supporting the roof of the fireboat berth. It had nearly fallen on the fireboat and trapped it.

In a few seconds the big engines on Fireboat 2 were roaring and it was on its way to the fire. When the fireboat approached the explosion area, the heat became intense. The pilot turned on fine-spray nozzles to keep the boat drenched with water. Still the paint on Fireboat 2's hull blistered and its windows cracked in the heat. The *Markay* was past saving. The crew of Fireboat 2 turned its attention to the warehouses. The boat had to pass through the blazing fuel to get to a spot from which it could reach the burning warehouses with its powerful streams of water. Using streams of water from its lower nozzles to sweep the fire away from the hull, the fireboat churned through.

Fireboat 2 was equipped for the job it was trying to do. It could pump 10,400 imperial gallons of water a minute. The boat carried an 11 centimeter (4½ in.) nozzle mounted on top of a water tower. There was another large nozzle mounted on the pilothouse, and several smaller nozzles were located around the sides. Some of these side nozzles could shoot water up at an angle to reach underneath a wharf. There was no danger of Fireboat 2's running out of water. Fireboats pump the very water they are floating on.

It wasn't long before Fireboat 2 was getting the best of the fire. When the sun came up over the smoky scene, the fire had been contained. The use of Fireboat 2 had cut the fire-fighting time by hours or days and saved millions of dollars' worth of property.

The fire departments of many cities like Los Angeles have added fireboats to help combat waterfront fires. Only the largest cities can afford them. Each one is built to fill the needs of a particular city at a cost of about one million dollars. Fireboats are seldom needed but they must be kept in top condition. When there is a big waterfront fire, nothing can substitute for a fireboat.

Fireboats are basically large tugboats equipped with fire-fighting equipment. They must have deep hulls to house the heavy pumps and other equipment they carry. At the same time, they must be of shallow draft to enable them to operate close to the docks. Usually the water tower is

Fireboats of the Los Angeles Fire Department. Notice the refineries and chemical plants in the background that make this a hazardous fire area

LOS ANGELES FIRE DEPARTMENT

collapsible, so it can be lowered when the fireboat goes under a low bridge. Fireboats must be fast, maneuverable, and seaworthy. They are sometimes required to leave the harbor to meet a burning ship steaming toward port.

Many fireboats are also equipped with foam-producing equipment. A liquid foaming agent is mixed into the water as it is shot out. This foam is used to smother burning petroleum floating on water.

Recently a new kind of fireboat, the "jet boat," has appeared on the waterfront. Although smaller than older fireboats, it is faster and more maneuverable. Jet nozzles shooting water rapidly to the rear provide the thrust to send the fireboat forward. Jet nozzles on the sides of the boat are used in maneuvering. They can be swiveled to move the boat backward, sideways, or around in a circle. When a conventional fireboat is pumping water, the force of the water leaving the nozzles causes a backward thrust. This steady push backward forces the boat away from the fire. For this reason, most fireboats have to use some of their nozzles to shoot water to the rear to offset the back thrust and keep the boat in place. Jet boats don't have to do this. They use some of their jet nozzles to keep them in one place while pumping water on a fire.

Fireboats are the most spectacular pieces of fire-fighting equipment around. When they go into battle with all their pumps sending out long arcs of water, they are something to watch.

SCUBA FIRE FIGHTING

On March 17, 1960, a welder was installing pipes under the new Matson Navigation Company's 55-million-dollar pier in Los Angeles harbor. A spark from the welding torch started a fire in the maze of creosoted wood under the pier. An alarm was sent in. Fire engines and fireboats rushed to the scene. But it took twelve hours to get the fire under control and twenty-four hours to put it out. The damage was estimated at three million dollars.

Fires under wharves and docks are extremely difficult to put out. The underside of waterfront structures is a maze of wood pilings, crosspieces, and planking. All this wood has been coated with creosote to preserve it. Creosote, being a petroleum product, is very flammable. When creosote burns, it gives off dense black smoke that irritates and blisters the skin. It can burn the throat and nose if inhaled.

Land-based firemen had trouble reaching the fire burning under the Matson wharf. They had to chop holes through the heavy timbers of the wharf and push spray nozzles down. But, since they couldn't see under the

Scuba fireman wearing his wet suit. Underwater techniques met much resistance at first

LOS ANGELES FIRE DEPARTMENT

wharf, the fire fighting had to be done by guess. A fireboat came in close to try to spray under the wharf. But many of the pockets of fire were so far back they couldn't be reached by the fireboat. The fire quickly spread under the wharves and warehouses. The fumes rolling up from the burning creosote blinded the fire fighters and made working conditions dangerous.

Assistant Chief W. W. Johnston, Jr., of the Los Angeles Fire Department, worked on the Matson fire. Chief Johnston, as well as other firemen at this fire, did scuba diving as a hobby. They were called upon to retrieve several pieces of equipment that had fallen into the water during the fire. While working under the piers, Chief Johnston realized that this fire could have been put out quickly in its early stages if there had been some way to fight it from under the wharves. Here, they could have spotted the places where the fire was burning and put them out. Why couldn't firemen wearing scuba-diving suits do this type of work?

At first most fire chiefs thought Johnston's idea silly and dangerous. How could scuba divers handle a twisting fire hose? Besides, they would be killed if they swam under a burning wharf in the heat and smoke of a creosote fire. It wouldn't work.

A scuba fireman with a supermonitor, a floating hose director
LOS ANGELES FIRE DEPARTMENT

Scuba firemen fighting a wharf fire. Burning timbers under the wharf cannot be reached from above

LOS ANGELES FIRE DEPARTMENT

But Chief Johnston began experimenting with his scuba diving idea in his spare time. There were many scuba divers in the Los Angeles Fire Department and some of them worked with Chief Johnston during their off time. They soon found that the fire hoses floated, even when filled with water, but the metal couplings and nozzles sank. One man could handle a 6.25-centimeter (2½ in.) fire hose in the water, although it took three firemen to handle it on land. The firemen found that heat reached 1000° C. (1900° F.) directly under the burning planking on a wharf. But the heat at the surface of the water only 2.4 meters (8 ft.) away was 40° C. (105° F.). Wet suits protected the firemen from the drips of the melting creosote. The men could also protect themselves with a cooling spray of water or submerge if it got too hot. If the smoke got too dense, they used a face mask and air tanks.

By this time Chief Johnston had convinced the fire department officials that they should give his idea a try. Other scuba-diving firemen heard about the unit and volunteered to work in it. Soon he had hundreds of volunteers putting in off-duty hours learning scuba fire fighting. Rigid tests of a fireman's ability to handle himself underwater reduced this number. Two hundred divers passed all tests and completed the training. Those selected could swim long distances carrying heavy weights on their backs. They could float for hours if necessary. The buddy system used by all scuba divers was used here too. Two men always worked together and watched out for each other.

The firemen set out to find answers to some of their problems. What could they do to make the metal couplings and nozzles float? At first plastic bottles were tied to them to keep them afloat. Later, firemen built floats, or monitors, to hold them up. Originally, the firemen paid for these improvements themselves and furnished their own diving equipment. Today, although scuba firemen get the same pay as other firemen, their diving equipment is furnished.

The divers kept working on improvements and eventually perfected a supermonitor made of polyurethane foam. These supermonitors are painted bright yellow to make them easy to see when the firemen are working in the water. They are about the size of half a surfboard. A nozzle in the center of the monitor can shoot out 360 imperial gallons of water a minute. The divers hold handgrips on the sides as they propel the board through the water. This can be done either swimming with flippers or the nozzle can be turned to shoot water to the rear to propel the monitor forward. One fireboat can pump the water for the fire hoses fastened to ten monitors.

When a call goes out for scuba firemen, a fireboat takes the divers to the fire. The firemen change into their wet suits during the trip. They smear their faces and hands with thick grease for protection from the creosote smoke. First, the fireboat tries to knock down some of the flames with large streams of water. Firemen on land are, at the same time, attacking the fire from above. When the flames are reduced, the scuba divers slip over the side and pay out their fire hoses. Each diver paddles under the wharf holding onto his monitor. If necessary, a spray of water is allowed to shoot from the nozzle in the center to keep the diver cool. Underneath the wharves, the firemen search out pockets of fire and use the spray of water to extinguish them.

The idea of scuba divers to help firemen has now spread to waterfront cities all over the world. Chief Johnston gets many calls from fire departments asking for information about his scuba-diving firemen. These diving firemen are also used in search-and-rescue operations. They often recover sunken boats or cars. They have been used to recover drowning victims or valuable equipment lost in the water.

In 1961, a fire alarm went out from Logan International Airport in Boston. A large passenger airplane had crashed. The airport and city firemen raced to the scene, but found that the plane had crashed short of the runway in the Boston harbor. A new scuba fire-fighting unit had just been organized in the Boston Fire Department. A call was sent out for all scuba firemen to report at once to the crash site. Fireboats were used as a base for the divers while they worked tirelessly in freezing weather. They recovered all bodies and brought up the parts of the plane necessary to find the cause of the crash. Although new to this type of work, the firemen got praise for their great performance.

7
CHEMICAL, ELECTRICAL, AND PETROLEUM FIRES

CHEMICAL FIRES

On April 16, 1947, the *Grandcamp* was moored at a dock in Texas City, Texas, being loaded with ammonium nitrate. This compound, widely used as a fertilizer, is very explosive. A small fire broke out in the hold of the *Grandcamp* and an alarm was turned in. The firemen arrived and boarded the ship. About this time the flames reached either the ammonium nitrate or the gas being released from it. The ship exploded with a terrific blast that could be felt miles away. Four hundred people, including the firemen, were killed in this one blast.

The explosion touched off a fire in the nearby Monsanto Chemical Plant, which had manufactured the ammonium nitrate. The chemicals stored in this huge plant burned fiercely. Two more ships being loaded with chemicals at nearby docks caught fire. The chemicals, already loaded onto the ships, could not expand when they got hot. So these two ships also exploded and set fire to eight oil tanks on shore.

The presence of such huge amounts of chemicals in one area made this Texas City disaster the worst chemical fire in history. It is estimated that 1,300 metric tons of ammonium nitrate were destroyed that day. About 1,000 people were killed and 4,000 injured. The destruction was so complete it was impossible to tell how many had died. Property loss was estimated to be over 67 million dollars.

The cheapest way to transport bulky chemicals is by water. So, many of the worst chemical fires have been on ships. One ship may contain tons of a certain chemical. Any time a large amount of chemicals is stored in one place, there is danger.

No ship carries enough fire-fighting equipment on board to handle a

Fire following the 1947 explosion at Texas City. Nearly 1,000 persons were killed, more than 3,000 injured
UNITED PRESS INTERNATIONAL

large fire. Even when trained firemen can be called, it is hard for them to get a fire line down a ship's narrow stairs and halls to the fire. The seat of the fire is difficult to find and reach. It is often buried deep in the hold under the cargo.

Chemical as well as petroleum fires start with a big bang. But firemen fear chemical fires above all others. In many chemical compounds, some atoms split off to unite with others—the compound decomposes. Decomposition produces heat and expansion. If the compound decomposes rapidly, it produces large volumes of gas—or an explosion.

Many chemicals behave in an orderly manner most of the time. But if they get mixed with certain other chemicals, there may be a violent reaction. Some of our most useful chemicals must be handled with great care. Different chemicals react to heat in different ways. Some explode; some give off poisonous fumes. Other fires require an outside source of oxygen,

Chemical and petroleum fires are often accompanied by explosions

LOS ANGELES FIRE DEPARTMENT

but many chemical formulas contain oxygen.

Most manufacturing plants use chemicals in their industrial processes. Because of the seriousness of chemical fires, fire departments often have a division trained in understanding chemical reactions. They study ways to fight chemical fires. But sometimes no one is sure what chemical is involved.

A fire in the Tarrant Building, a drug warehouse in New York City, was this type of fire. Every kind of chemical used in making drugs was stored there. As the containers became hot and broke during the fire, all the drugs were mixed together. Some chemicals exploded; others gave off poisonous fumes. Firemen couldn't get near the fire. It had to be fought from outside the building and upwind.

When it is too dangerous to enter a building, firemen use high-expansion foam to fight the fire. A foam generator is rolled as close to the fire as possible. Then a long "sock," a sort of overgrown fire hose, is pushed into the building, and the foam generator is turned on. A room can be filled with foam to cut off oxygen from the fire. Within a few hours the foam will gradually break down and disappear. It does no damage to the contents of the room.

ELECTRICAL FIRES

When Fred got home from school, his mother wasn't home. Fred didn't care. He flipped a switch and the lights came on. He flipped another switch and the television was on. It was a little cold in the house so he flipped a switch and the heat came on. Fred was always hungry after football practice, so he poured some popcorn and oil into the popcorn popper and flipped another switch. Then he opened the refrigerator. The smoothly running machine kept the milk cold—just the way Fred liked it. Soon Fred was sitting in front of the television, warm and comfortable with warm popcorn to eat and cold milk to drink.

The next day when Fred came home, he started flipping switches again. Nothing happened. There was no electricity. So Fred had no lights, no heat, no television, no popcorn, and the milk was warm. Fred was astonished to find out how much his family depended on electricity.

Today we expect electricity to be a trouble-free helper. We use it for so many things, we forget that electricity can be dangerous. Most of the houses we live in were built when there were far fewer electrical appliances than we use today. The wiring in many houses isn't adequate for present needs.

When we use several electrical appliances, we sometimes try to draw more electricity than the wires can handle. A safety device is built into every house to prevent the wires from being overloaded—the fuses. The fuses put into a house when it is built are designed for the wiring in that house at that time. They should always be replaced with fuses of the same size. If the fuses are replaced with larger ones to handle more electricity, the house wires heat up and may cause a fire. Electrical fires often occur because of overloaded wires or electricity escaping through a broken or loose wire. All electrical appliances should be kept free from flammable dust and lint. Motors, which require lubrication, should be regularly cleaned and oiled to keep them from overheating. If an appliance is used for work that is too difficult for it, the wires overheat. Not taking care of, or misusing electrical equipment is one of the main causes of fires.

Electricity can start a fire in wood, paper, and other building materials. Water can be used to put out this type of fire if the electricity is shut off first.

Fires that occur in an electrical appliance pose a special problem. Since water conducts electricity, it cannot be used to extinguish this type of fire. Electrical fires are extinguished with carbon dioxide gas or other dry or liquid chemicals. These agents cut off the oxygen, and the fire quickly dies out.

Electricity makes countless jobs easier for us. But electricity is rather like a caged tiger. We must always treat it with respect.

PETROLEUM FIRES

Osaga is a small village near Osaga Lake. For nine months of the year Osaga is a sleepy little village. But in June the summer visitors begin to arrive and summer homes are unboarded. Once a week a gasoline tanker rumbles into town bringing gasoline for the visitors' cars and powerboats. One Saturday morning the tanker was a little late as it came down the last hill. The driver did not realize how fast he was traveling until he tried to take the sharp turn leading into town.

He had a sinking feeling as he turned the wheel sharply to the left to avoid the cars parked on the narrow street. The heavily loaded trailer shuddered as it tried to follow the cab around. Then it toppled over, pulling the rest of the rig onto its side. Everyone in town heard the crash. One seam of the tank split open and thousands of gallons of gasoline began to spill out into the street.

The pedestrians and shopkeepers ran for their lives when they saw the cause of the crash. One spark here would start a holocaust. The firemen at

A petroleum fire is smothered with foam from twin turrets on a special truck

UNITED PRESS INTERNATIONAL

the firehouse only a block away quickly began hooking up a fire hose. The fire chief felt his best chance of avoiding a disastrous fire was to wash the gasoline quickly into the storm drains. Streams of water were directed on the wrecked tanker and its dangerous cargo. But just as this plan seemed to be working—whoooooooooosh! The gasoline vapors had found a spark and the gasoline burst into flames. The firemen now turned their hoses on the stores in an effort to save them.

By now the water was carrying the blazing gasoline down the gutters. Cars parked at the curbs were set on fire as the burning gasoline passed

A fog mist nozzle is being used to cool a petroleum fire

UNITED PRESS INTERNATIONAL

under them. The gutters carried the fire into the storm drain without doing much damage there. Then the flaming gasoline passed out of the storm drain to float down a stream lined with summer homes. Many of these houses, or their verandas, were built out over the water. As the burning gasoline traveled downstream, it passed under many of the houses, setting them on fire.

This fire is an example of the problems involved in fighting petroleum fires. Petroleum is lighter than water, so it floats on top when the two are mixed. The firemen of Osaga had hoped to wash the gasoline out of town before it could catch fire. But the fact that gasoline and its vapors are very flammable prevented that.

Every liquid vaporizes, or gives off a gas, at certain temperatures. The temperature is different for each liquid. Gasoline starts to vaporize at 43° C. (45° F.) below zero. Gasoline vapor, being heavier than air, settles down and spreads out. A single spark will cause it to ignite. Some flammable gasses cannot be smelled. Others, such as gasoline vapor, can be smelled only for a short time. After the first two minutes of exposure, a person cannot smell gasoline fumes and may forget they are there.

When gasoline vapor is enclosed, it has an explosive force ten times greater than that of dynamite. It takes eighty-three sticks of dynamite to equal the explosive force of vapor from one gallon of gasoline.

Great care must be used in handling, manufacturing, and storing petroleum products. Oil refineries are located in an area away from other buildings. Each tank farm, where large quantities of petroleum are stored, has its own fire-fighting force. The oil tanks are surrounded by dikes. If the contents of one tank catch fire, the dikes keep the fire from spreading. Since lightning is an important cause of fires in oil tanks, each tank is equipped with a lightning rod. Tank covers are tightly sealed to keep the dangerous vapors from escaping.

Foam is the best agent for fighting petroleum fires. It cuts oxygen off and so smothers the fire. Foam is produced in several ways. One method is to use a hopperlike device called a foam generator. Foam powder is added to water in the hopper to form a foam much like shaving lather. A large hose, or sock, is used to carry the foam to the fire. Another method is adding liquid foam to a stream of water. This foam expands to cover one thousand times the area that could be covered by the same amount of water alone.

Sometimes the area of burning oil is too large to be blanketed by foam. Then a hose with a fog mist nozzle is used. A water mist cools the fire down and helps keep nearby structures from bursting into flames. A stream of water is never used.

Small oil fires around a home can be extinguished with the carbon dioxide found in many fire extinguishers. Since this gas weighs one and one half times more than air, it sinks down to smother a fire. If no fire extinguisher is available, bicarbonate of soda, commonly called baking soda, will put out a small oil fire. This is the compound used in many commercial fire extinguishers.

8
AIRPORT FIRES

FIRE HAZARDS AT AIRPORTS

"Tower to Flight 95. Tower to Flight 95. You are cleared to taxi to runway three. Wait there for further instructions. Takeoff time will be around 10:05. There are five airplanes ahead of you."

Every day instructions like these are given by controllers from towers at thousands of airports in the United States. More than 10,000 airplanes are estimated to be in the air at all times during the day. Thousands of commercial flights take off and land every day. At a busy airport pilots must await their turn.

Every precaution is taken to make air travel safe. Usually it is. But now and then something goes wrong and there is an accident. All aircraft carry flammable fuel. A single commercial jet carries thousands of gallons. In an airplane accident, fuel tanks are often broken and the heat of the engines sets the fuel on fire. If smoke and flames envelop the aircraft, the people inside can survive for only four minutes. If the fuselage is broken, and this usually happens in a crash, the survival time is much less—only a few seconds. Airports must have a quick method of reaching and extinguishing aircraft fires.

Any type of fire at an airport can spread quickly to other structures. The aviation fuel stored there makes the area especially hazardous. The land surrounding each large city airport has developed into a small city. Aircraft industries, cargo handling terminals, and facilities for repairing and maintaining aircraft are all built close to the airport. Hundreds of airplanes may be parked there at one time. Quick action in case of fire is necessary to prevent a disaster.

Large airports are located some distance from their cities. It would take

Hazards of airport fires: one fireman, with feet on fire, runs for the protection of foam. Men in the background are extinguishing flames on the coat of a fireman

UNITED PRESS INTERNATIONAL

Airport firemen, wearing fire-resistant suits, pour foam on a burning aircraft

UNITED PRESS INTERNATIONAL

from fifteen minutes to half an hour for the city fire department to respond to a call from the airport. That time could mean the difference between life and death in an airplane crash. Airports must maintain their own fire-fighting departments.

AIRPORT FIRE-FIGHTING EQUIPMENT

A large airport usually needs from two to ten pieces of fire-fighting equipment. The airport has its own firehouse located on the edge of a main runway.

City fire departments get their calls from a centrally located fire alarm system. Airport firemen get their alarm calls from the tower or from the pilot of an incoming craft. A hot line goes from the control tower to the airport firehouse. When this telephone is picked up in the control tower, alarm bells begin clanging in the firehouse. The fire crew is on the crash trucks and ready to roll by the time the captain puts the phone down. These firemen are especially trained in methods of fighting aircraft fires.

Equipment used by airport fire fighters differs in many ways from other fire-fighting equipment. An airplane fire could happen anyplace. Fire mains cannot carry water to every place on the runway where it might be needed. Airport fire engines carry their own water and other fire-fighting materials. Because of the fuel aircraft carry, water alone is never applied to a burning plane.

Airport crash trucks are not painted the traditional fire-engine red. Chrome yellow, which shows up better from the air, is the color most often used. Pilots coming in for an emergency landing try to stop their planes close to the crash trucks.

Airport firemen do not wear the traditional fire coats and hats. Their clothes must resist the high temperatures of burning fuel. The coats are made of silver-colored, heat-reflective cloth. A fireman also wears a hood of the same material with a heat-resistant face shield. Boots are flameproof and aluminized.

Some firemen use the fire rescue, or "hot papa," suits that enable them to pass safely through flames. Such a suit will resist temperatures of up to 982° C. (1800° F.). Gas and smoke masks are worn under the hot papa suits.

Airport crash trucks come in more than one size. Most airports have speedy lightweight crash trucks of up to 3,600 kilograms (8,000 lbs.). Their purpose is to reach the crash site and start rescue operations within seconds.

They carry enough foam and other extinguishing agents to fight a fire for five minutes. The heavy crash wagons (up to 27 metric tons) are designed to reach the crash site and get into position before the lighter trucks have used up their supply of foam.

All crash trucks have large oversized wheels for travel over rough ground. Sometimes an airplane crashes short of the runway. The crash rig has several nozzles mounted on top of the truck. Other nozzles are mounted on the front bumper. These nozzles are directed toward the ground and use foam to sweep the fire away from the advancing crash truck. Some nozzles, or guns as they are called, can shoot a stream of foam 60 meters (200 feet).

The large crash trucks carry several thousand gallons of water and hundreds of gallons of concentrated liquid foam. The foam-making process is started while the rig is still racing to the fire. The foam is mixed with water and pumped out under high pressure. One gallon of water and foam concentrate mixed together expands to ten gallons of foam.

Dry chemical powders are also carried by the crash wagons. They too are shot out under pressure. Since dry chemicals do not have to be mixed, they are often used until enough foam can be generated to smother a fire.

A tanker, or "nurse" truck, often accompanies the crash wagons to a fire. This rig carries additional water and foam concentrate to resupply the crash wagons. Sometimes it too is equipped with nozzles to spray foam.

A rescue rig is part of the fire-fighting equipment of most airports. It carries rescue equipment such as floodlights, metal saws, and portable stairways to help get people out of the aircraft.

When fighting a hot fire, airport firemen can operate the nozzles and other fire-fighting equipment from inside the enclosed cab for protection. A hatch in the roof of the cab enables a fireman to stand inside the cab while operating equipment. Only the top part of the fire fighter's body is exposed to the heat. He may be enclosed in a fire-resistant suit and helmet.

As the operator of the crash wagon approaches the burning aircraft, he looks the situation over. He positions the truck so that the firemen can fight the fire with the wind behind them. The streams of foam travel farther if they are thrown in the same direction the wind is blowing. In addition, the flames and smoke are blown away from the firemen. As the firemen move up, they use the low nozzles on the front bumpers of the crash truck to sweep the fire before them. At the same time, they blanket the fuselage with foam. Then firemen in hot papa suits try to open an escape route to rescue people trapped in the aircraft.

CRASH LANDINGS

"Flight 71 to tower. Flight 71 to tower."

"This is the tower. Come in, Flight 71."

"We seem to be having trouble with our landing gear. The instruments indicate that it has not descended."

"This is the tower. Go back up to 5,000 feet. We will send someone up to look you over."

This is one of the emergencies that may make a crash landing necessary. An inspection of the underside of the aircraft may show that the landing gear is in place. A faulty message has been flashed on the instrument panel. But if the landing gear will not descend, a crash landing must be made.

The pilot circles the field to use up his aviation fuel. He only retains

Fireman using foam to smother fire in airplane
UNITED PRESS INTERNATIONAL

enough for the landing. While he is doing this, the crash crew covers the runway with foam. All flights are rerouted to other runways. The foam helps reduce friction and prevents sparks as the plane slides in for a belly landing. The instant the plane is on the ground, the crash rigs race to it, ready to pour on more foam at the first sign of fire.

Another aircraft emergency is a fire in an engine. The crash rigs stand ready to smother the fire with carbon dioxide or dry chemicals when the plane lands. Foam or water would damage an expensive jet engine.

9
TWO CONFLAGRATIONS

CHICAGO—1871

Did the great fire in Chicago start when Mrs. O'Leary's cow kicked a lantern over in her barn? We probably will never know. Fact and fancy often get mixed when there is a great disaster. But this is the way the story has come down to us.

In 1871, Chicago had a well-organized fire department. It was one of the first fire departments to replace volunteers with paid employees. On October 8, 1871, the fire department was taking a well-earned rest. It had been fighting the worst fire in Chicago's short history.

Chicago had a fire problem that summer of 1871. The city had grown rapidly, and hastily constructed buildings crowded the shores of Lake Michigan. Many were of wood with tar-paper roofs. That summer was one of the driest on record. There had been no rain since July and water was scarce. One fire had followed another all summer.

In this last big fire, thirty firemen had been injured and were now off work. The fire department had lost several pieces of equipment and thousands of feet of valuable hose. The weary firemen hoped for a period without fires so that they could rest up and get their equipment and men working again. But everyone was worried. The combination of heat and high winds had made conditions in Chicago ripe for a disastrous conflagration.

De Koven Street was a few blocks from the smoking debris of the first fire. In spite of this fire, a big party was going on there at the McLaughlins' that Sunday evening. Everyone was drinking Mr. McLaughlin's famous milk punch, until they ran out of milk. Mr. McLaughlin went to ask Mrs. O'Leary, his landlady, if she could get them some more milk. She kept a cow in the barn behind her house. Although the cow had already been milked, kind

Mrs. O'Leary consented to go out in back and see if she could get more milk. She took a kerosine lantern and a pail and went out to the barn. The cow wasn't very happy about being bothered again when she had just settled down for the night. Mrs. O'Leary set the lantern and the pail down behind the cow. The annoyed cow gave a kick and sent the lantern and pail flying into a pile of hay. The kerosine spilled over the hay and exploded into flames. In a few minutes the fire had raced up the stack of hay and had ignited the dry barn roof. Soon the strong winds were blowing smoke,

"The Great Fire at Chicago, Oct. 8, 1871." Currier & Ives lithograph

THE HARRY T. PETERS COLLECTION,
MUSEUM OF THE CITY OF NEW YORK

flames, and embers onto the roofs of the surrounding buildings.

Someone ran to a nearby drugstore to put in a fire alarm from the box there. But during the recent big fire, the fire alarm system had been damaged. The alarm never reached the fire department. Fifteen or twenty minutes later, a fireman at the firehouse noticed the flames, which were already lighting the sky.

The fire department had to guess where the fire was located. It guessed wrong and valuable time was lost while the firemen looked for it. By the time they arrived, the fire had been burning for almost forty-five minutes. Fire Chief Williams sent in a second alarm. After the reinforcements arrived, the fire department seemed, at first, to be getting the best of the fire. But then the high wind took over. It blew the streams of water back toward the firemen and kept the air full of flying firebrands and embers. The fire quickly spread from one building to another.

Chief Williams tried to outguess the fire. He moved his tired fighters to a position in the path of the fire to try to make a stand. But the wind blew the fire over the firemen's heads to start in a new place. Sometimes the wind changed and the fire started off in a new direction. As the fire spread, the firemen thought surely they could stop it at the south branch of the Chicago River. But the windblown fire easily jumped the river. The flames were too hot for the firemen to get in close with the hoses. No matter how much the steam fire engines bellowed and belched, the streams of water could not reach the heart of the fire.

By then the entire Chicago Fire Department was fighting the fire. Around midnight Chief Williams sent out a call for help. A telegram was sent to every city and town within fifty miles, asking their fire departments to send help.

The gasworks was lying in the path of the spreading fire. To try to avoid a huge explosion there, the workers released the gas into the city's sewers. This caused another problem. The escaping gas spread through the sewers and came up in places far from the site of the fire. Any spark there caused the gas to catch fire and spread more fires.

The streets were jammed with refugees trying to escape from the burning city. Most of the refugees were struggling with carts of household goods, cows, horses, and children. The streets were so clogged that the firemen couldn't move their fire equipment from one place to another. Fire engines and valuable hose were lost to the advancing fire.

By 3:30 on the morning of October 9, the fire had reached the waterworks. When this building burned, the pumps that were supplying the fire mains were put out of commission. Now there was no water with which to fight the fire.

At daylight, fire companies began arriving from nearby cities and towns. Some fire departments traveled as far as 480 kilometers (300 mi.) to help Chicago. They brought much-needed equipment and hoses. But all were hampered by lack of water. Even with all of this help, a fire line could not be established to stop the spreading fire.

The battle went on all day. Thousands of terrified refugees had fled to Lincoln Park on Michigan Lake. The fire burned down to the open space of the park.

On the morning of October 10, the weary firemen had been battling the fire for more than thirty-six hours without rest. It was completely out of control. There seemed to be no way to stop the fire until all of Chicago had burned down.

In those early morning hours, the tired firefighters couldn't see the towering thunderheads that were building up over the stricken city. No one could hear the crash of thunder over the crackle of the flames. No one saw the lightning through the smoke that covered the city. It began to rain. Only a few drops fell at first, then it began to pour. In a short time the blocks of burned debris and ashes began to sizzle. The fires burning over one third of Chicago began to die out. The firemen had not been able to put out the fire, but the rain did.

By the evening of the third day of the fire, the people of Chicago began to count up their losses. Over one third of the city had been destroyed. Eighteen thousand buildings had been burned. At least two hundred people were dead, but there was no way to count them all. Fifty thousand people left town, convinced that Chicago would never recover from the disaster.

But Chicago did recover. Within three years, Chicago had been rebuilt and there was no trace of the terrible fire. But the people of Chicago had learned a lesson. This time the city was built as fireproof as possible.

SAN FRANCISCO—1906

The crust of our earth is made up of huge blocks, or plates. In some areas, one huge block is slowly moving in one direction while the block next to it is moving in the opposite direction. As the blocks try to move past each other, they are held back by friction. Tension between two blocks builds up until, at last—pow! The two blocks slide past each other with a jolt—an earthquake occurs. Cities built on or near the boundary of two blocks, known as a fault, feel many earthquakes.

San Francisco is built a few miles from the world's most famous fault —the San Andreas Fault. One block there is moving southeast, while the other is moving northwest at the rate of two inches a year. In 1906 no earthquakes had occurred in this area for some time. People there were beginning to feel very secure. But tensions were building up along the San Andreas Fault. Suddenly, in the early morning of April 18, 1906, the two sides of the fault slid past each other with a severe jolt. In some places the two blocks moved, in relation to each other, as much as 6.3 meters (21 ft.).

Let's look at San Francisco in 1906. It had been growing by leaps and bounds since the gold rush days of 1849. Since the city is built on a peninsula, building space was always scarce. Many buildings, at that time, were built on man-made land. That is, land that had been reclaimed from the bay by covering mud flats with rubbish and gravel. An earthquake causes this type of land to shake like jello. The wooden buildings of San Francisco were crowded together. Since there wasn't room to spread out, many buildings were high. These conditions caused San Francisco to have serious fire problems. But the people weren't worried. They had a superb fire department, one of the best fire departments in the nation.

Late on the night of April 17, 1906, Engine Company Five of the San Francisco Fire Department returned from another troublesome fire. The chief of the fire department, Dennis Sullivan, supervised the men as they cleaned and put the equipment and horses away. Then Chief Sullivan went home. He didn't have far to go. The Sullivans lived at the firehouse in an apartment on the top floor of the three-story building. When Chief Sullivan was on duty, Mrs. Sullivan slept in the front bedroom while her husband slept in the back bedroom. She didn't like to be awakened by the fire gong ringing beside the bed at all hours of the night. When Chief Sullivan crawled into bed that April night, he hoped for some well-earned rest. Maybe there would be no more fires tonight.

A little after five on the morning of April 18, Chief Sullivan awoke with a start. He was sure he was having a nightmare. He thought someone was shaking him violently. The chief sat up and saw the walls of his room swaying back and forth. Pictures were falling off the walls and his shaving mug fell from the shelf with a crash. Then he heard a thundering roar. It seemed to come from the front of the apartment. He jumped out of bed and ran to his wife's room. When he opened the door, he again had the feeling that this was a nightmare. In the half-light of dawn, he could look across the street. Looking up, he could see the sky getting lighter in the east. The roof and wall of the bedroom were gone. Dust was so thick, Chief Sullivan couldn't see into it. He tried to grope his way across the room to where his wife's bed stood. But he took only a couple of steps before he felt himself falling. The floor too was gone.

Meanwhile, the firemen on the second floor were also awakened by the earthquake. They slid down the pole and tried to quiet the terrified horses. The firemen knew that fires would follow the earthquake. They began to dig the fire engines out from under the rubble that covered them. Imagine their surprise—almost as soon as they started digging, they found

The San Francisco fire, April 18, 1906. View from Nob Hill looking downtown

WELLS FARGO BANK HISTORY ROOM, SAN FRANCISCO

their own fire chief. He had been seriously injured, but told them what had happened.

The firemen managed to get a buggy and horse out to take Chief Sullivan to the hospital. It took longer to find and dig out Mrs. Sullivan. But she was lucky. When she had fallen, her mattress had landed on top of her. The mattress had trapped air for her to breathe, and it had prevented injury from the debris that fell on top of her. Chief Sullivan died of his injuries.

By then fire companies all over town were digging their steam engines out and trying to get them ready to roll. But some of the fire equipment was seriously damaged. Firemen and fire horses had been killed or injured. The fire alarm system all over town had been damaged by the earthquake. No calls came in to the firehouses. But the columns of smoke rising all over the city told the men of the fire department that they were needed. One by one the fire departments got their horses out, harnessed them to the fire engines and rolled toward the fires. Many detours had to be made around the masonry clogging the streets.

The firemen struggled up the steep hills to the fires. They got their fire lines laid, then made a terrifying discovery. The earthquake had broken almost every water main in the city.

The fire department had no way to fight the fires that were spreading rapidly. The firemen retreated to the edge of the bay. There they laid thousands of feet of hose and started pumping water from the bay. They would try to save the downtown area where most of the fires had started. Market Street, which runs through the downtown area, is a wide street. The fire department thought it could make a stand there and keep the fire from crossing Market Street. But then the wind came up and, with a shortage of water, the firemen were unable to hold the fire at Market Street.

The fire department was working at a terrible disadvantage. It was short of men, equipment, horses, and water. The fire by now had spread across Market Street and was working its way uphill. The fire department gathered up its equipment and moved up to Van Ness Street. This is another wide street that bisects San Francisco. Here it prepared to make another stand.

The firemen were afraid they had no more chance of making a stand at Van Ness than they had had at Market Street. All of the city between Van Ness and the bay was now burning. Every now and then an explosion rocked the city as explosive material was set off by the fire. People retreated up the hills away from the advancing fire. All that day and the next the fight went on at Van Ness Street. The Army and Navy joined the fight. Buildings were dynamited to clear open spaces to stop the advance of the fire. Patrols were organized to shoot anyone found looting. Time after time the wind blew the fire across Van Ness Street, but each time the fires were put out before they could spread. In spite of all their losses and the shortage of water, the San Francisco Fire Department stopped the fire at Van Ness.

The fight went on for three and one half days before all the fires in San Francisco were out. The fire did much more damage than the earthquake. On the fourth day the people began to count up their losses. Five hundred people were killed. Many more buildings were lost there than were burned in the great Chicago fire. But an accurate count was hard to make. Ten square kilometers (4 sq. mi.) of the city lay in ruins. Much of it was just rubble. All over the rest of the city, there was damage from both the earthquake and the fires that followed. Insurance claims for 350 million dollars were filed. But help poured in to San Francisco from all over the world. Within a few years a new city had arisen by the bay. But San Francisco too had learned a lesson. This time the city was built as fireproof and earthquake proof as possible.

10
FIRE PREVENTION

FIRE SAFETY EDUCATION

Have you ever heard of Smokey, the bear, or Sparky, the fire dog? Most young people in the United States today have seen pictures of these two animals and know what they stand for. They are mascots for two different fire prevention programs. These are only two of many such programs. Their aim is to educate people, especially children in elementary schools, in ways to prevent fires. Most such programs stress home fire safety. But the one run by the United States Forest Service stresses forest fire prevention.

The Junior Fire Department, one typical fire safety education program, was started in 1953 by the Los Angeles Fire Department's chief. In this program children are enlisted as junior fire fighters. They are educated in fire safety through talks, movies, and printed material. They win promotions by studying their official manuals and passing tests.

Why do fire departments spend their time teaching children and adults about fire safety? Don't they have enough work to do putting out fires? The idea of the kind of work a fireman should do has changed. A few years ago all fire departments worked only on fighting fires after they had once started. Fire chiefs were chosen for their ability to fight fires. Today, the emphasis is on preventing fires. Many big-city fire chiefs are chosen because of their work in fire prevention. Today's fire departments spend part of each day on preventive work.

Firemen feel that the best time to learn about fire prevention is when one is young. This education should make a person more conscious of fire hazards for the rest of his life. Junior fire prevention plans turn out boys and girls who watch their own homes for conditions that could start a fire. They know what to do in case of fire. Children have a big stake in fire safety. Half of all fire deaths occur in the home. Of these, one out of three is a child.

This can happen to any home! Every family should work out an escape plan

LOS ANGELES FIRE DEPARTMENT

IN CASE OF FIRE

A person who has studied fire safety knows what to do if a fire should break out in the house. Do you? The first thing is to sound a warning to alert others in the house. Young children who lack this training sometimes hide themselves in a closet or under the bed when there is a fire. These are the two places firemen look first for children who have been trapped by a fire. Often when they find them, it is too late.

What would a junior fire fighter do upon awaking in the night and thinking there was a fire in the house? A trained person does not run to the door and throw it open. One breath of superheated air can kill. If a person understands fire safety, he feels the door first to see if it is hot. Then he looks around the edges for signs of seeping smoke. If things look safe so far, he opens the door a crack, but stays behind it, ready to slam it shut if heat and smoke rush in.

In case there is a fire, the trained person knows it is usually safer to go out the window. If the window doesn't open easily, he breaks it with a chair or anything heavy. Windows don't cost much and getting out in a hurry may save a life. It is safe to jump from a first- or second-story window. But a trained person doesn't try to jump from the third floor or higher. At that height, he straddles the windowsill and screams for help. But one leg is kept inside the room and wrapped in something heavy to protect it from heat. When a junior fire fighter gets out of the house, he checks on others in the family by going to their windows and breaking them if necessary.

A safety-conscious family works out an escape plan to use in case of fire. The entire family knows what to do and arranges to meet in a certain place. Then the members will know if someone is trapped inside. Many fire-related deaths have been caused by a person's going back into a blazing building to look for someone who is already safely outside.

A trained person knows the number of the firehouse in case it is necessary to call the fire department. In a situation where it is best to stay out of the house, he knows where the nearest fire alarm box is located.

BUILDING FOR FIRE SAFETY

Most big conflagrations have occurred in places where little attention was given to fire safety during building. At one time structures were all built as inexpensively as possible. Little thought was given to making them fireproof. Now that has changed. Most cities have adopted fire codes requiring buildings to be built to certain fire-resistant standards. Engineers study ways to make buildings fireproof. Usually a fire wall, a thick wall made of unburnable material, is required between buildings. Fire codes are especially strict for public buildings such as theaters, and concert and dance halls. Curtains and other furnishings in these buildings must be made of nonflammable material.

Fire prevention does not stop with buildings. Engineers do research trying to make household furnishings, toys, appliances, and clothes more fire-resistant. They study ways to decrease the fire hazards in many of the things we use every day. They educate people to the dangers in many common things. One example is the spray can. Many people don't realize that the gas propellants in spray cans will burn like a blowtorch if they are exposed to an open flame. They will explode if thrown into a fire. Environmentalists are working hard to banish these cans.

Special safety devices are required in public and large office buildings. Fire alarms and fire drills are often required so that people know the best way to evacuate a building. Many buildings are required to have automatic sprinkler systems that turn on when the temperature in a room reaches a certain point. Research has shown sprinklers to be one of the most effective safety measures. Another safety device is an alarm that is set off when smoke or heat is detected in a room. Fire extinguishers are sometimes required to be placed in every room, and the people who work in the rooms should be taught how to operate them.

ENFORCING FIRE LAWS

Sometimes shortsighted people feel they should not be required to build according to a fire code. These codes make building more expensive. Owners may feel that a fire in their home or office is their own business. After all, no one will be hurt but themselves.

This is far from true. A fire in one building can quickly spread to others. No one can think only of themselves when it comes to fire safety. Fire safety is everyone's business.

Fire safety inspectors, usually firemen, inspect all buildings to see if they are built and operated according to the fire code. They pay special attention to buildings where people gather. They check to see if fire escapes, fire doors, and other safety devices are in good condition. Surprise checks are made. Fire inspectors look to see whether the fire exits are locked, whether halls and aisles are kept clear. When they find violations, citations are issued. These citations can result in a fine or imprisonment. Buildings can be closed down until the violations are corrected. Thousands of lives and millions of dollars in property are saved every year because these laws are strictly enforced.

Firemen have learned that it is better to prevent a fire than to put one out after it has started. For that reason they spend time on fire safety education, fire safety inspections, and other fire law enforcement work. Firemen today would be happy if they never had to roll out the fire engines. Then they could spend all their time stopping fires before they start.

11
FIRE FIGHTERS TODAY

THE CHANGING FIREHOUSE

What if a twentieth-century fireman could be a modern Rip Van Winkle? Suppose this fireman slept for twenty years as did Van Winkle and then woke up? Would he notice any differences around the firehouse? If a new firehouse had been built in the same location while he slept, would it look the same? Would the firemen there be wearing the same type of clothing and would they still be using the same equipment?

Firehouses built today bear little resemblance to firehouses built twenty years ago. Our Rip Van Winkle would probably pass by a modern firehouse without knowing what it was. Old firehouses were usually dark places with two or more floors. They were hot in summer and cold in winter. A tall drying tower for the fire hoses stood out in back. The firehouse contained a large room with card and billiard tables so the firemen could entertain themselves. There were usually several of the town idlers at the firehouse. They often dropped by to play cards and chat with the firemen.

The firemen slept upstairs and came sliding down the pole when the fire alarm clanged. Many injuries were caused when people stumbled through the hole for the fire pole.

Firehouses built today usually have only one floor. Several small firehouses at different locations are used rather than one large firehouse in the center of town. The new firehouses are light, airy, and attractive. Heating and cooling systems keep the firemen comfortable. The old drying tower no longer stands out in back. Now the coiled fire hoses are usually dried in a large oven at low temperatures. Card and billiard tables have been replaced by television. The firemen are too busy to chat with casual visitors. Schoolchildren on field trips are the most frequent visitors to firehouses.

The clanging fire gong no longer awakens the firemen in the middle of the night. Studies have shown that these loud bells put a strain on the firemen's ears and nerves. They have been replaced by a subdued buzzer or chime.

FIRE CLOTHES TODAY

The style of fire clothes worn today would look familiar to our modern Rip Van Winkle. The cut of fire clothes has not changed much in the last one hundred years. The type of fire clothing first adopted has proved to be the most practical and offers the most protection. But today some clothes are made of different materials.

The fireman's turnout coat is waterproof and fire-resistant. A warm lining is snapped in during cold weather and waterproof pants are worn. During a fire, firemen are drenched with spray from the hoses. In freezing temperatures the water turns to ice. A fireman would freeze if he were wet through to the skin. The turnout coat is worn summer or winter, since it protects the fireman from the heat of the fire. The older fire coats are very heavy, for they were made of black rubberized material. New fire coats are made of lightweight flame-resistant material. Some are aluminized to reflect heat. The new coats are often orange, which shows up best at night.

The distinctive hats worn by firemen today are almost the same as they were two hundred years ago. The crown is still shaped like an eagle's head. The hat is hard to protect the fireman's head from falling debris. The broad back brim has proved most effective in keeping water and embers from sliding down inside the turnout coat. It can be turned to the front to be used as an emergency heat shield. Early helmets were made of hard leather. Today they are made of impact-resistant plastic or metal. Many now have a heavy transparent heat shield for the face that can be raised or lowered.

Over the front brim of the fire helmet is a flat place—often called the shield. This spot contains information to help identify the fireman. The number of the fire company appears in large numerals—often with the name or initials of the fire department. Under this is the fireman's own number. The color of the helmet also gives information. Although most firemen wear black helmets, the chief wears a white one to make him easier to locate during a fire. All officers have their helmets marked in a distinctive fashion. It may be only a white shield in front or a white line running over the black helmet. In many fire departments, the truck company men have red shields on their helmets; the rescue company men, blue shields.

The fireman's badge today, like the first ones, is shaped like the Maltese cross worn by knights on crusades. The four arms of the cross are wide and flaring to symbolize the spreading wings of a bird protecting its young. A phoenix, a mythical bird that is supposed to arise from ashes, is often pictured on a fire badge. The badge of each company shows the special tools it uses. Those of aerial firemen show ladders and pike poles; those of an engine company show crossed nozzles. The chief's badge is often topped by an eagle, the ruler of the birds.

THE FIREMAN'S DAY

It is morning at the firehouse. Firemen are polishing and repairing the fire engines. Hoses are being rolled into neat doughnuts. Floors are being mopped and, if there is a fire pole, that is being polished. Everyone is busy and the firehouse buzzes with activity. Then, as if a spell had been cast, each man stands motionless and silent. A few lips move as each fireman counts.

Buzz . . . Buzz . . . Buzz. The firemen are counting the number of times the buzzer sounds. Is this call for them? Firemen get so used to counting buzzers or bells that they count any sound they hear.

"Everything goes! Everything goes!" yells the dispatcher. Now the firehouse bursts into a frenzy of activity. It may look like chaos, but each man knows what he is doing. All the firemen race for the fire engines. If it is an older firehouse, the firemen upstairs come sliding down the pole. They grab their fire coats and helmets before leaping for the fire trucks. The motor firemen jump into the driver's seats and the powerful motors roar into life. The tillerman belts himself into the lonely seat on the back end of the ladder truck. Someone hits the button and the huge front doors roll up.

Within seconds the fire engines are careening down the street—lights flashing, sirens screaming. The swaying engines race through stop lights and around slower-moving cars as the drivers try to get out of their way.

When the trucks pull up in front of a smoking apartment house, the men are on the ground tugging at the hoses before the wheels have stopped. Each man has a certain job to do. Some pull the heavy black hose off the truck and couple it to the fire hydrants. Others pull the canvas hoses off the trucks and lay the fire lines. The motor firemen adjust the valves, and, within minutes, the pumper is throwing out a steady stream of water. The entry men have taken their tools and are opening locked doors. The ladder company officers and crew are already searching the building for people trapped or overcome by smoke. Others are raising the ladders so they can enter the upper stories. The roof men are making their way to the roof to "vent" the fire. They open the roof to release dangerous trapped smoke and heat.

A fireman's life is filled with danger and excitement. It is not the average eight-to-five job. When a fireman is on a fire, he stays until the fire is out. No one walks off the job because it is quitting time. A fire company lives and trains together until the firemen work as a team. Nothing is left to chance. If one man doesn't do his job, valuable property is lost—maybe a

Firemen are always ready to rush to someone's aid in any emergency

LOS ANGELES FIRE DEPARTMENT

family loses its home. Even worse, the fire may be fatal to someone—often a fireman.

Anyone who watches firemen fighting a fire is impressed by the dedication of these men. They risk their lives every time they go to work. Fire fighting is one of the most dangerous of all occupations.

Only a small portion of a fireman's time, though, is spent actually fighting fires. What do firemen do when they aren't fighting a fire?

The hours on duty for most firemen have not changed since the days of the first paid fireman. They are on duty for twenty-four hours at the firehouse and are then off for twenty-four hours. Some people have criticized the hours firemen work, since they get paid for the entire twenty-four hours they spend at the firehouse. This includes the time they are sleeping. But more firemen would be needed to man the firehouse if this system were not used.

The same group of men lives and works together for twenty-four hours at a time. Strong friendships develop among firemen as they fight fires together. When another person's life depends on you and your life depends on him, you learn to trust him.

At most firehouses the day starts at 8:00 A.M. with a lineup. The firemen coming on duty for their 24-hour shift line up. The firemen who have just finished their 24-hour shift sometimes line up too. At this time the company chief can check for absences and tell the firemen about special problems and tasks to be done.

The day starts with firemen giving the firehouse and its equipment the same careful attention that sailors give their ship. Everything must be clean and in perfect working order. Each day everything is cleaned and polished. After this, while the firemen are still wearing their work clothes, training drills and exercises are held. Firemen practice until they can do all types of fire-fighting activities automatically. The firemen work at a drill tower, practicing raising ladders, carrying hoses up them, rescuing people and sliding down a rope lifeline. Driving drills help familiarize motor firemen with the streets and driving hazards.

In the afternoon, classes are held several times a week. An officer, with the aid of a blackboard, works on special problems. Maps are used to make the firemen more familiar with their fire area. Sometimes a drill is held when the officer calls out addresses and the firemen respond by telling the quickest way to reach each address. Firemen memorize locations, such as schools and hospitals, where helpless people could be trapped by fire. They learn the locations of places that have special fire hazards, such as oil refineries or lumber mills. Each day they review any obstacle, such as street repairs, that would cause a traffic obstruction.

In the afternoon firemen may work on fire prevention. They make calls in residential or business areas to inspect for, and warn people of, fire hazards on their property. Sometimes they conduct classes on fire safety.

The fireman's day is not all work. Firemen must keep in good physical condition. Active games such as handball help to keep them that way.

Firemen take turns cooking nourishing meals at the firehouse. Most firemen are good cooks. If they can't cook to start with, the complaints of other firemen soon make them improve their skills. Several fire departments have even published cookbooks of their own recipes.

But in spite of all the improvements, today's firemen work under the same tensions firemen have always worked under. They have the same tradition of quick service. Firemen are never out of reach of some type of alarm. Day or night, working or playing, eating or sleeping, a fireman who hears an alarm will drop everything and rush to someone's aid. Firemen are available in case of a fire, a heart attack, or even to rescue a cat that is afraid to climb down from a treetop. A fireman can never really relax. He must always be ready to answer a call for help.

BECOMING A FIREMAN

Volunteers in the early fire departments raced to fires because they enjoyed the excitement of answering fire calls and liked the social life of the fire department. Most volunteers then had little training and no tradition of helping others. They often fought among themselves as each company jostled to be first at a fire. When the firehouses first began to be manned by paid firemen, political bosses handed out these jobs to campaign workers. No attention was given to education, physical condition, or aptitude for the job.

Today that has changed. Candidates must have at least a high school diploma before they are allowed to take the written and oral civil service examinations for fireman. Many fire departments require graduation from a junior college with a major in fire science.

Candidates must also pass rigid physical and medical examinations. Applicants are screened to see whether they are able to live and work closely with others. They are tested to see whether they have a fear of high places. About half of the applicants are weeded out during these tests.

Because of the tough physical requirements of fire fighting, it is usually only the young men who can pass the tests. A few women serve as fire fighters, but mostly as volunteers in small towns. The strenuous work of fire fighting may go on for hours. Firemen must climb, often pulling heavy hoses, up ladders and through smoke-filled buildings in heats of over 38° C. (100° F.). This work requires a strong person in top physical condition. Several hours of fire fighting require as much physical energy as a week of most other kinds of work. Older firemen have usually worked up to supervisory positions or work that is less physically demanding.

Beginning firemen who have passed all the tests go through a three-month training period. Fire recruits are drilled almost like army recruits. They go through a tough session of physical fitness and performance drills. Many can't take it and drop out. But the firemen who are left acquire the same tradition of teamwork and pride in their organization found in the armed services. As the rooky fireman trains, he gains confidence in himself. He finds he is able to do things he never thought possible. He learns to carry a man down a 30 meter (100 ft.) ladder and rappel himself down a five-story building. He learns to think of himself as a fireman—someone who is looked to by others whenever a crisis occurs. He learns to keep his head and think coolly in an emergency.

But a fireman's education does not stop when he gets a job. All firemen must learn how to give advanced first aid. Then they must take refresher courses to keep up with new methods. Those assigned to rescue units may be called upon to treat heart attack and drowning victims and respond to other medical emergencies. They may be called upon to deliver a baby if the mother can't reach a hospital. Serious injuries often occur at fires and every fireman learns how to treat the injured.

Firemen may take part of their training through college classes. Many earn college degrees while working. A knowledge of mathematics, physics, and chemistry is necessary in some branches of fire work. Advanced degrees are usually required for all chief officers. In large fire departments, the department head may have earned a doctorate in fire science. These degrees are usually acquired while the men are working for the fire department. Most of the cost is absorbed by the department's training budget.

Motor firemen must have a thorough knowledge of automotive and other mechanical equipment. They take classes to improve their mechanical skills. Motor firemen are expected to do the repair work on fire engines and all the other mechanical equipment they use. Firemen today are much better educated than they were twenty years ago.

The pay for firemen is good when all the benefits, such as pensions, are considered. Their salaries compare favorably with those paid in private industries. A fire chief in a large city may make as much as $35,000 a year and many firemen make $15,000 a year.

But the job of fireman has its drawbacks. Fire fighting is dangerous. The chance of a fireman's being killed on the job is thirteen times as high as it is for other workers. Every week in the United States an average of one fireman is killed and four hundred others are injured. Firemen serve because there are few other jobs that offer them the satisfaction or excitement of fire fighting. Every ring of the alarm brings a new challenge and a chance to help others. Firemen risk their lives because they live by the tradition of helping those in trouble.

GLOSSARY

Aerial ladder An extensible ladder mounted on a truck.
Air tanker A tank-equipped bomber used to drop water and chemicals on a fire.
Atom The smallest particle of an element.
Azimuth indicator A device used to find a compass direction.
Borate bomber A tank-equipped aircraft used to drop water and chemicals, often borate, on a fire.
Brush breaker A rural fire truck equipped with a water tank and a shield to break down brush.
Bucket brigade A line of people formed to pass buckets of water quickly to extinguish a fire.
Buddy system The pairing of two divers to look out for each other.
Canvas hose The white flat hose used to carry water from the pumper to the fire.
Chaparral Heat- and drought-resistant small trees and brush found on western hills.
Conflagration A huge fire covering a large area.
Creosote A petroleum product used to preserve wood.
Decomposition The process by which a substance is broken down into separate elements.
Fault In geology, the earthquake-prone meeting place of two of the earth's plates.
Fire hook An iron hook on a rope used to pull burning thatch from a roof.
Fire main A duct or pipe designed to carry water for fire extinguishing.
Fire triangle The three things needed for combustion—oxygen, fuel, heat.
Fire wall A fireproof wall.
Friction The rubbing of two things together.

Gooseneck crane A hinged boom equipped with water pipes and nozzles to fight fires on upper floors.

Hand tub An early fire-fighting device used for pumping water.

Hotshot crew Ground fire fighters for forest fires.

Inhalator A device to help a person inhale air.

Jumpmaster The supervisor of a parachute jump.

Kindling temperature The lowest temperature at which a substance takes fire and burns.

Life net A hand-held net used to catch jumping people.

Molecule A particle of substance composed of one or more atoms.

Monitor (In scuba fire fighting) A float with a centrally mounted nozzle that can be turned in any direction.

Motor fireman The fireman who drives, operates, and repairs the motorized equipment.

Oxidation The process by which oxygen unites with another element.

Oxygen A colorless, odorless gas.

Pack pump A small water tank and pump used to extinguish small brush and grass fires. It is usually carried on the back.

Phlogiston An element thought, until the 1700's, to cause material to burn.

Pulaski A combination curved hoe and ax used in fighting forest fires.

Rescue units A specially designed truck carrying equipment and personnel to give first aid or rescue people.

Resuscitator A machine that forces oxygen into the lungs of a person suffering from lack of air.

San Andreas Fault The earthquake-prone meeting place of two of the earth's plates—located in California.

Santa Ana wind. A strong northeast wind that blows periodically in Southern California.

Scuba fire fighter A fireman who wears scuba-diving gear to fight dock fires.

Smoke jumper A fire fighter who parachutes to forest fires.

Snorkel A hinged boom equipped with water pipes and nozzles used to fight fires on upper floors.

Spontaneous combustion Rapid oxidation without external heat that produces heat and can cause fire.

Spotter airplane An airplane used for making an aerial examination of the ground.

Standpipe A large water pipe built into skyscrapers to carry water to the upper floors.

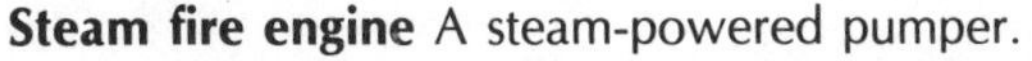

Steam fire engine A steam-powered pumper.

Suction hose The heavy black hose used to pump water from a fire hydrant, or other source, to the pumper.

Superpumper An enormous pumper that pumps water from a harbor, river, etc.

Supertender A hose truck equipped with water cannons that accompanies Superpumper.

Vaporize Change into gas.

Volunteer fire department A fire department staffed by unpaid firemen.

Volunteer fireman An unpaid fireman.

Water cannon A nozzle that shoots powerful streams of water under high pressure.

Water engine A tub with a nozzle from which water could be squirted under pressure.

Water tower A nozzle-equipped tower used to throw a stream of water to upper floors.

Wet suit The rubber suit worn by a scuba diver in cold water.

BIBLIOGRAPHY

BOOKS

Aitken, Frank W., and Hilton, Edward, *A History of the Earthquake and Fire in San Francisco.* Ed. Hilton Co., 1906.

Colby, C. B., *Smoke Eaters.* Coward-McCann, Inc., 1954.

Cromie, Robert, *Great Chicago Fire.* McGraw-Hill Book Co., Inc., 1958.

Da Costa, Phil, *Fire Fighting Apparatus.* Bonanza Books, 1964.

Ditzel, Paul C., *Firefighting: A New Look in the Old Firehouse.* Van Nostrand-Reinhold Company, 1969.

Dougherty, Thomas F., and Kearney, Paul W., *Fire.* G. P. Putnam's Sons, Inc., 1931.

Haywood, Charles F., *General Alarm.* Dodd, Mead & Company, Inc., 1967.

Hill, Charles T., *Fighting a Fire.* Century, 1916.

Holden, Raymond, *All About Fire.* Random House, Inc., 1964.

Holzman, Robert S., *Romance of Firefighting.* Harper & Brothers, 1956.

Kenoyer, Natlee, *The Firehorses of San Francisco.* Westernlore Press, 1970.

King, William T., *History of the American Steam Fire-engine.* Owen Davies Publisher, 1960.

Shannon, Terry, and Payzant, Charles, *Smokejumpers and Fire Divers.* Golden Gate Junior Books, 1969.

Tamarin, Alfred H., *Fire Fighting in America.* The Macmillan Company, 1971.

Wilson, Dorothy, *Fire Prevention.* Franklin Watts, Inc., 1965.

MAGAZINE AND NEWSPAPER ARTICLES, BOOKLETS

Bentley, Clyde, "Smokejumping: Only a Few Can," *Redding Record-Searchlight,* May 23, 1973.

Fire Control. California State Department of Education, 1974.

First Water. American LaFrance Corporation, 1972.

Greenburg, Dan, "The Perils of Battling Smoke and Flame," *Life,* March 24, 1972.

Lawson, John, "Fighting Forest Fires No Easy Job," *Redding Record-Searchlight,* Oct. 9, 1971.

"Lifeline," *The Naval Safety Journal,* Sept. and Oct. 1976.

Malcolm, Andrew H., "The New Breed of Men Who Jump Into Fires," *San Francisco Examiner & Chronicle,* Sept. 15, 1974.

Olney, Ross and Pat, "The Devil Winds and the Mountain Fires," *The Elks Magazine,* Oct. 1974.

Shelby, Clifford, "Rugged Training Readies Smokejumpers," *Redding Record-Searchlight,* July 16, 1968.

Solomon, Mark, "Jumping Into the Fire," *San Francisco Examiner & Chronicle,* Aug. 21, 1971.

U.S. Department of Agriculture, *Air Attack on Forest Fires.* U.S. Government Printing Office, 1960.

INDEX